T0183262

SpringerBriefs in Mathematics

Series Editors

Nicola Bellomo, Torino, Italy

Michele Benzi, Pisa, Italy

Palle Jorgensen, Iowa, USA

Tatsien Li, Shanghai, China

Roderick Melnik, Waterloo, Canada

Otmar Scherzer, Linz, Austria

Benjamin Steinberg, New York, USA

Lothar Reichel, Kent, USA

Yuri Tschinkel, New York, USA

George Yin, Detroit, USA

Ping Zhang, Kalamazoo, USA

SpringerBriefs present concise summaries of cutting-edge research and practical applications across a wide spectrum of fields. Featuring compact volumes of 50 to 125 pages, the series covers a range of content from professional to academic. Briefs are characterized by fast, global electronic dissemination, standard publishing contracts, standardized manuscript preparation and formatting guidelines, and expedited production schedules.

Typical topics might include:
A timely report of state-of-the art techniques A bridge between new research results, as published in journal articles, and a contextual literature review A snapshot of a hot or emerging topic An in-depth case study A presentation of core concepts that students must understand in order to make independent contributions

SpringerBriefs in Mathematics showcases expositions in all areas of mathematics and applied mathematics. Manuscripts presenting new results or a single new result in a classical field, new field, or an emerging topic, applications, or bridges between new results and already published works, are encouraged. The series is intended for mathematicians and applied mathematicians. All works are peer-reviewed to meet the highest standards of scientific literature.

Titles from this series are indexed by Scopus, Web of Science, Mathematical Reviews, and zbMATH.

Marius Ghergu

Partial Differential Inequalities with Nonlinear Convolution Terms

Springer

Marius Ghergu (iD)
School of Mathematics and Statistics
University College Dublin
Dublin, Ireland

ISSN 2191-8198   ISSN 2191-8201   (electronic)
SpringerBriefs in Mathematics
ISBN 978-3-031-21855-2   ISBN 978-3-031-21856-9   (eBook)
https://doi.org/10.1007/978-3-031-21856-9

© The Author(s), under exclusive license to Springer Nature Switzerland AG 2022
This work is subject to copyright. All rights are solely and exclusively licensed by the Publisher, whether
the whole or part of the material is concerned, specifically the rights of translation, reprinting, reuse
of illustrations, recitation, broadcasting, reproduction on microfilms or in any other physical way, and
transmission or information storage and retrieval, electronic adaptation, computer software, or by similar
or dissimilar methodology now known or hereafter developed.
The use of general descriptive names, registered names, trademarks, service marks, etc. in this publication
does not imply, even in the absence of a specific statement, that such names are exempt from the relevant
protective laws and regulations and therefore free for general use.
The publisher, the authors, and the editors are safe to assume that the advice and information in this book
are believed to be true and accurate at the date of publication. Neither the publisher nor the authors or
the editors give a warranty, expressed or implied, with respect to the material contained herein or for any
errors or omissions that may have been made. The publisher remains neutral with regard to jurisdictional
claims in published maps and institutional affiliations.

This Springer imprint is published by the registered company Springer Nature Switzerland AG
The registered company address is: Gewerbestrasse 11, 6330 Cham, Switzerland

Preface

This brief research monograph aims to investigate old topics with contemporary mathematical methods. Indeed, partial differential equations (in short PDEs) with convolution terms have been around for nearly a century, being introduced by D. Hartree in 1928 in connection with the Schrödinger equation in quantum physics. Despite this fact, the first mathematical studies towards understanding these objects only emerged in the mid-1970s.

In their full generality, these equations display a non-local structure. Classical methods such as maximum principle or sub- and super-solution method do not apply to this context. There is a breadth of literature that has developed in the last decade around variational methods to study PDEs with convolution terms. However, most of these techniques do not apply to our context here, because the goal of the book is to discuss partial differential inequalities (instead of differential equations) for which there is no variational setting. Therefore, the tools available to study differential inequalities featuring nonlinear convolution terms are apparently very limited.

This current work is an offshoot of the author's original research on the subject during the last few years and brings forward other methods that prove to be useful in understanding the concept of a solution and its asymptotic behaviour related to partial differential inequalities with nonlinear convolution terms. It promotes and illustrates the use of a priori estimates, Harnack inequalities and integral representation of solutions.

The first chapter of the book presents the motivation for this research direction as well as the differences and challenges that arise in the study of differential equations versus differential inequalities. Chapter 2 investigates partial differential inequalities for quasilinear elliptic operators driven by the m-Laplace or the mean curvature operator. We distinguish between inequalities posed on bounded and exterior domains. In our approach, a crucial role is played by various integral a priori estimates. The study is then continued in Chap. 3 in case of punctured balls where a priori estimates are employed alongside with Harnack inequalities. Chapter 4 discusses differential inequalities for the polyharmonic operator $(-\Delta)^m$. In such a setting, we bring in a new ingredient which proves to be effective to our purpose: integral representation of solutions. Such a technique is peculiar to potential analysis and may yield undoubtedly benefits in the qualitative study of PDEs as it is illustrated in Chap. 4. Chapter 5 investigates parabolic inequalities involving quasilinear operators. This chapter may be regarded as the time-dependent continuation of the inequalities discussed in Chap. 2. New types of a priori estimates are obtained to derive the nonexistence of weak solutions. Chapter 6 deals with higher order evolution equations – for both time and space variables – with convolution terms.

Each chapter defines specific methods of approach and concludes with further discussions on the topic. In this way the reader is provided with a general overview on the literature pertaining to the subject. Two appendices are included which contain properties of superharmonic functions and necessary background on Harnack inequalities for quasilinear elliptic equations. The prerequisites for this monograph lie at the level of a graduate course in PDE drawing from basic knowledge on Sobolev spaces, distributions, classical and weak solutions for linear PDEs.

The present work is a self-contained study on the topic and appeals to graduate and postgraduate students as well as to researchers in the field of partial differential equations and nonlinear analysis.

Dublin, Ireland   Marius Ghergu
July 2022

Contents

Chapter 1
Preliminary Facts

1.1 Motivation

This work is devoted to the study of the differential inequalities of the form:

$$\pm \mathscr{L}u \geq (K * u^p)u^q \quad \text{in } \Omega \subset \mathbb{R}^N, \, N \geq 1, \tag{1.1}$$

and their time-dependent counterpart:

$$\frac{\partial^k u}{\partial t^k} \pm \mathscr{L}u \geq (K * u^p)u^q \quad \text{in } \mathbb{R}^N \times (0, \infty), \, k \geq 1, \tag{1.2}$$

where $\mathscr{L}$ is a differential operator having one of the following forms:

- $\mathscr{L}$ is a quasilinear operator in divergence form $\mathscr{L}u = \text{div}[\mathscr{A}(x, u, \nabla u)]$ where $\mathscr{A}$ satisfies some structural assumptions;
- or $\mathscr{L}$ is a higher-order differential operator given by the iteration of the Laplace operator $\mathscr{L}u = (-\Delta)^m u, \, m \geq 1$.

The potential $K \in L^1_{loc}(\mathbb{R}^N)$ is assumed to be positive and $K * u^p$ denotes the standard convolution operator:

$$(K * u^p)(x) = \int_\Omega K(x - y)u^p(y)dy \quad \text{for all } x \in \Omega. \tag{1.3}$$

The prototype model is $K(x) = |x|^{-\alpha}$ for $\alpha > 0$ which has several physical motivations as we explain in the following; note that condition $K \in L^1_{loc}(\mathbb{R}^N)$ yields $\alpha \in (0, N)$. Another particular case for K is the *Riesz potential* I_α of order $\alpha \subset (0, N)$ given by $I_\alpha : \mathbb{R}^N \setminus \{0\} \to \mathbb{R}, \, I_\alpha(x) = A_\alpha |x|^{\alpha - N}$ where $A_\alpha > 0$ is a specific constant that depends on N and α which is defined in terms of Gamma function Γ as follows:

© The Author(s), under exclusive license to Springer Nature Switzerland AG 2022
M. Ghergu, *Partial Differential Inequalities with Nonlinear Convolution Terms*,
SpringerBriefs in Mathematics, https://doi.org/10.1007/978-3-031-21856-9_1

$$A_\alpha = \frac{\Gamma\left(\frac{N-\alpha}{2}\right)}{\Gamma\left(\frac{\alpha}{2}\right)\pi^{N/2}2^\alpha} > 0.$$

One specific feature of the Riesz potential is the so-called semigroup property ,
namely,

$$I_\alpha * I_\beta = I_{\alpha+\beta} \quad \text{for all } \alpha, \beta \in (0, N), \alpha + \beta < N.$$

The study of convolution terms in time-dependent partial differential equations
goes back to near a century ago. Indeed, the equation

$$i\psi_t + \Delta\psi + (|x|^{-\alpha} * \psi^2) = \psi \quad \text{in } \mathbb{R}^N \times (0, \infty), N \geq 1, \tag{1.4}$$

was introduced by D.H. Hartree [Har28a, Har28b, Har28c] in 1928 for $N = 3$ and
$\alpha = 1$ in relation to the Schrödinger equation in quantum physics.

Looking for *standing wave solutions* of (1.4) in the form $\psi(x, t) = e^{it}u(x, y)$,
we are led to

$$-\Delta u + u = (|x|^{-\alpha} * |u|^2)u \quad \text{in } \mathbb{R}^N, \tag{1.5}$$

which bears the name *Choquard* or *Choquard-Pekar equation*.

For $N = 3$ and $\alpha = 1$, Eq. (1.5) was introduced in 1954 by S.I. Pekar [Pek54]
as a model in quantum theory of a Polaron at rest (see also [DA10]). In 1976, P.
Choquard used (1.5) in a certain approximation to Hartree-Fock theory of one-
component plasma (see [Lie76]). In 1996, Eq. (1.5) appears in a different context,
being employed by R. Penrose [Pen96] as a model of self-gravitating matter (see,
e.g. [Jon95, MPT98]) and it is known in this context as the *Schrödinger-Newton
equation*.

The first mathematical study of (1.5) are due to E.H. Lieb [Lie76] and P.-L.
Lions [Lio80, Lio84] starting from the mid-1970s. Let us point that (1.5) and more
generally, equation

$$-\Delta u + u = (|x|^{-\alpha} * |u|^p)|u|^{p-2}u \quad \text{in } \mathbb{R}^N \tag{1.6}$$

have a variational structure. Indeed, solutions of (1.6) are seen as minimisers of the
energy functional:

$$\mathcal{E}(u) = \frac{1}{2}\int_{\mathbb{R}^N}\left(|\nabla u|^2 + u^2\right) - \frac{1}{2p}\int_{\mathbb{R}^N}\int_{\mathbb{R}^N}\frac{|u(x)|^p|u(y)|^p}{|x-y|^\alpha}dxdy.$$

By the Hardy-Littlewood-Sobolev inequality, the above quantity is finite for all
$u \in W^{1,2}(\mathbb{R}^N) \cap L^{\frac{2Np}{2N-\alpha}}(\mathbb{R}^N)$. In 2013 V. Moroz and J. Van Schaftingen [MV13a]
obtained that (1.6) has nontrivial (i.e. non-identical zero) solutions if and only if

$$2 - \frac{\alpha}{N} < p < 2 + \frac{4 - \alpha}{N - 2}. \tag{1.7}$$

Moreover, if p satisfies (1.7), then (1.6) admits ground state solutions, that is, solutions $u \in W^{1,2}(\mathbb{R}^N) \cap L^{\frac{2Np}{2N-\alpha}}(\mathbb{R}^N)$, $u \neq 0$ such that

$$\mathscr{E}(u) = \inf \left\{ \mathscr{E}(v); v \in W^{1,2}(\mathbb{R}^N) \setminus \{0\} \text{ and } \int_{\mathbb{R}^N} \left(|\nabla v|^2 + v^2 \right) = \int_{\mathbb{R}^N} \left(|x|^{-\alpha} * |v|^p \right) |v|^p \right\}.$$

1.2 Equations Versus Inequalities

The study of elliptic inequalities in unbounded domains goes back to the early 1980s although elliptic equations in $\mathbb{R}^N$ have been discussed, for radially symmetric solutions, at least one century ago by Emden [Emd07] and Fowler [Fow14, Fow20].

Gidas and Spruck obtained in the celebrated paper [GS81] that the semilinear equation

$$-\Delta u = u^p \quad \text{in } \mathbb{R}^N, \, N \geq 3, \tag{1.8}$$

has no C^2-solutions for $1 \leq p < \frac{N+2}{N-2}$ and that the upper exponent $\frac{N+2}{N-2}$ is sharp. Instead, if one considers the related inequality

$$-\Delta u \geq u^p \quad \text{in } \mathbb{R}^N, \, N \geq 3, \tag{1.9}$$

then the optimal range for nonexistence changes to $1 \leq p < \frac{N}{N-2}$ and this new upper exponent $\frac{N}{N-2}$ is also sharp (see, e.g. [BCN94, MP01]). Since then, such results have been extended to many differential operators. For instance, the authors in [CDM08] discuss Liouville-type results for

$$\mathscr{L}u = -\text{div}[\mathscr{A}(x, u, \nabla u)] \geq f(u) \quad \text{in } \Omega \subset \mathbb{R}^N.$$

The approach in [CDM08] relies essentially on representation formulae for linear inequalities, nonlinear capacity methods and the weak form of the Harnack inequality. In [BM98] the authors use a blow-up argument to derive a priori bounds for solutions and thus to obtain Liouville-type results for inequalities and their corresponding systems. Other types of problems may be found in [BFP15, KLM05, KLZ03, LLM07, MP01].

A systematic study of the inequality

$$\mathscr{L}u = -\text{div}[\mathscr{A}(x, u, \nabla u)] \geq |x|^\sigma u^q \quad \text{in } \Omega,$$

along with the corresponding system

$$\begin{cases} -\operatorname{div}[\mathscr{A}(x, u, \nabla u)] \geq |x|^a u^p v^q \\ -\operatorname{div}[\mathscr{B}(x, v, \nabla v)] \geq |x|^b u^r v^s \end{cases} \quad \text{in } \Omega,$$

is carried out in [BP01] for various domains $\Omega \subset \mathbb{R}^N$, such as open balls and their complements, half balls and half spaces.

The quasilinear elliptic inequality

$$\operatorname{div}(A(|\nabla u|)\nabla u) \geq f(u) \quad \text{in } \mathbb{R}^N,$$

is discussed in [PRS07, PSZ99] in connection with the strong maximum principle and the compact support principle. More recently, quasilinear elliptic inequalities and systems integrate the gradient term in the nonlinearity: the authors in [FPR10] discuss the quasilinear coercive inequality

$$\operatorname{div}(g(x)|\nabla u|^{p-2}\nabla u) \geq h(x)f(u)\ell(|\nabla u|) \quad \text{in } \mathbb{R}^N.$$

To the best of our knowledge, the first results dealing with quasilinear elliptic inequalities featuring nonlocal terms appear in [CMP08] where local estimates and Liouville-type results are obtained for

$$-\operatorname{div}[\mathscr{A}(x, u, \nabla u)] \geq K * u^q \quad \text{in } \mathbb{R}^N,$$

where $\mathscr{A}$ is S-m-C, $K \in L^1_{loc}(\mathbb{R}^N)$, $K \geq 0$ and $q > 0$.

The contrast between the study of the local equation (1.8) and the local inequality (1.9) as we illustrated above extends as well to the study of nonlocal equations and inequalities. Indeed, more recently Moroz and Van Schaftingen [MV13b] obtained that the semilinear elliptic inequality

$$-\Delta u \geq (|x|^{-\alpha} * u^p)u^q \quad \text{in } \mathbb{R}^N \setminus \overline{B}_1 \qquad (1.10)$$

has positive solutions if an only if:

- $p + q > \dfrac{2N - \alpha}{N - 2}$

- $\begin{cases} \min\{p, q\} > \dfrac{N - \alpha}{N - 2} & \text{if } 0 < \alpha < 2, \\ p > 1, q \geq 1 & \text{if } \alpha = 2, \\ p > \dfrac{N - \alpha}{N - 2}, q > 1 - \dfrac{\alpha - 2}{N}p & \text{if } 2 < \alpha < N. \end{cases}$

We shall see that the same result holds if the inequality (1.10) is posed in $\mathbb{R}^N$. Now, comparing the above results for the nonlocal inequality (1.10) with those in (1.7) for the nonlocal equation (1.6), we see that (1.10) has positive solutions for any $p, q > 1$ sufficiently large, while (1.6) has positive solutions only if p lies in a bounded open interval given by (1.7).

On the other hand, there is no gap between the existence of solutions to equations and inequalities in a semilinear parabolic setting. Precisely, Fujita [Fuj66] and Hayakawa [Hay73] proved that the problem

$$\begin{cases} \dfrac{\partial u}{\partial t} - \Delta u = u^p & \text{in } \mathbb{R}^N \times (0, \infty), \, p > 1, \\ u(x, 0) = u_0(x) & \text{in } \mathbb{R}^N, \end{cases} \tag{1.11}$$

has positive solutions for some data u_0 if and only if $p > 1 + \frac{2}{N}$. The same result holds if we replace the main equation in (1.11) with an inequality.

Chapter 2
Quasilinear Elliptic Inequalities with Convolution Terms

2.1 Introduction and Examples

In this chapter, we discuss the quasilinear elliptic inequality

$$\mathscr{L}u = -\text{div}[\mathscr{A}(x, u, \nabla u)] \geq (|x|^{-\alpha} * u^p)u^q \quad \text{in } \Omega, \tag{2.1}$$

where $\Omega \subset \mathbb{R}^N$, $N \geq 1$ has one of the following three shapes:

- Ω is open and bounded;
- Ω is the whole space $\mathbb{R}^N$;
- Ω is the exterior of a closed ball in $\mathbb{R}^N$.

In the last case, the inequality (2.1) will be considered in $\Omega = \mathbb{R}^N \setminus \overline{B}_1$, but the arguments we construct in the following are valid if Ω is the complement of any smooth and nondegenerate compact set. The quantity $|x|^{-\alpha} * u^p$ represents the convolution operation:

$$(|x|^{-\alpha} * u^p)(x) = \int_{B_1} |x - y|^{-\alpha} u^p(y) dy. \tag{2.2}$$

Throughout this chapter, $\mathscr{A} : \Omega \times [0, \infty) \times \mathbb{R}^N \to \mathbb{R}^N$ is a Caratheodory function, that is, $\mathscr{A}$ is measurable and $\mathscr{A}(x, u, \cdot) : \mathbb{R}^N \to \mathbb{R}^N$ is continuous for all $(x, u) \in \Omega \times [0, \infty)$. Further, $\mathscr{A}$ is assumed to fulfil one of the structural conditions below.

Definition 2.1 Let $m > 1$.

- We say that $\mathscr{A}$ is *weakly-m-coercive* (in short W-m-C) if

$$\mathscr{A}(x, u, \eta) \cdot \eta \geq C|\mathscr{A}(x, u, \eta)|^{m'} \quad \text{for all } (x, u, \eta) \in \Omega \times [0, \infty) \times \mathbb{R}^N,$$

© The Author(s), under exclusive license to Springer Nature Switzerland AG 2022
M. Ghergu, *Partial Differential Inequalities with Nonlinear Convolution Terms*,
SpringerBriefs in Mathematics, https://doi.org/10.1007/978-3-031-21856-9_2

where $C > 0$ is a constant and m' is the Hölder conjugate of $m > 1$, that is, $\frac{1}{m} + \frac{1}{m'} = 1$.

- We say that $\mathscr{A}$ is *strongly-m-coercive* (in short *S-m-C*) if

$$\mathscr{A}(x, u, \eta) \cdot \eta \geq C_1 |\eta|^m \geq C |\mathscr{A}(x, u, \eta)|^{m'} \quad \text{for all } (x, u, \eta) \in \Omega \times [0, \infty) \times \mathbb{R}^N,$$

where $C, C_1 > 0$ are constants.

- We say that $\mathscr{A}$ is of mean curvature type (H_m) if

$$\mathscr{A} = \mathscr{A}(\eta) : \mathbb{R}^N \to \mathbb{R}^N, \quad \mathscr{A}_i(\eta) = A(|\eta|)\eta_i,$$

where $A \in C[0, \infty) \cap C^1(0, \infty), t \longmapsto tA(t)$ is increasing and there exists $M > 1$ such that

$$\begin{cases} A(t) \leq Mt^{m-2} & \text{for all } t > 0, \\ A(t) \geq M^{-1}t^{m-2} & \text{for all } 0 < t < 1. \end{cases}$$

Example 2.2 Let $\mathscr{A} : \Omega \times [0, \infty) \times \mathbb{R}^N \to \mathbb{R}^N$ be defined by $\mathscr{A}_i(x, u, \eta) = \sum_{j=1}^{N} a_{ij}(x, u, \eta)\eta_i$. Then $\mathscr{A}$ is *W-m-C* provided that there exists $C > 0$ such that we have

$$\sum_{i,j=1}^{N} a_{ij}(x, u, \eta)\eta_i\eta_j \geq C \Big[\sum_{i=1}^{N} \Big(\sum_{j=1}^{N} a_{ij}(x, u, \eta)\eta_j \Big)^2 \Big]^{m'/2} \quad \text{for all } (x, u, \eta) \in \Omega \times [0, \infty) \times \mathbb{R}^N.$$

Example 2.3 The standard m-Laplace operator $\mathscr{A}(x, u, \eta) = |\eta|^{m-2}\eta$ is (H_m) and *S-m-C* for any $m > 1$.

Example 2.4 The m-mean curvature operator given by

$$\mathscr{A}(x, u, \eta) = \frac{|\eta|^{m-2}}{\sqrt{1 + |\eta|^m}} \eta \tag{2.3}$$

is *W-m-C* but not *S-m-C*. Also, $\mathscr{A}(x, u, \eta)$ is (H_m) provided that $m \geq 2$.

Example 2.5 Assume $\mathscr{A}_i(x, u, \eta) = A(x, u, |\eta|)\eta_i$, where $A : \Omega \times [0, \infty) \times [0, \infty) \to \mathbb{R}$. Then:

- $\mathscr{A}$ is *W-m-C* if there exists $M > 1$ such that

$$\begin{cases} 0 \leq A(x, u, t) \leq Mt^{m-2} & \text{for all } t > 0, \\ A(x, u, t) \geq M^{-1}t^{m-2} & \text{for all } 0 < t < 1. \end{cases}$$

- $\mathscr{A}$ is S-m-C if

$$M^{-1}t^{m-2} \leq A(x, u, t) \leq Mt^{m-2} \quad \text{for all } t > 0, \tag{2.4}$$

for some constant $M > 1$.

One should note that (H_m) operators satisfy the following comparison result.

Proposition 2.6 *Assume $\mathscr{A}$ satisfies (H_m) and $f : \mathbb{R} \to \mathbb{R}$ is a nonincreasing function. Let $\Omega \subset \mathbb{R}^N$ be a bounded domain and $u, v \in W^{1,1}(\Omega) \cap C(\overline{\Omega})$ satisfy*

$$\mathscr{L}u = -\text{div}\left[A(|\nabla u|)\nabla u\right] \leq f(u), \quad \mathscr{L}v = -\text{div}\left[A(|\nabla v|)\nabla v\right] \geq f(v) \quad \text{in } \Omega$$

and $u \geq v$ on $\partial\Omega$. Then $u \geq v$ in Ω.

Proof Assume the above conclusion does not hold. Thus, one may find $\varepsilon > 0$ such that the set $\Omega_\varepsilon = \{x \in \Omega : u(x) < v(x) - \varepsilon\}$ is nonempty. Clearly, $\overline{\Omega}_\varepsilon \subset \Omega$ and by continuity arguments, one has $u = v - \varepsilon$ on $\partial\Omega$.

Let $\phi : \mathbb{R} \to [0, \infty)$ be a nondecreasing function such that $\phi \equiv 0$ on $(-\infty, 0]$ and $\phi' > 0$ on $(0, \infty)$. Since f is nondecreasing, we have

$$\int_{\Omega_\varepsilon} \left(\mathscr{L}v - \mathscr{L}u\right)\phi(v - \varepsilon - u)dx \geq \int_{\Omega_\varepsilon} (f(v) - f(u))\phi(v - \varepsilon - u)dx \geq 0.$$

By the divergence theorem , this yields

$$\int_{\Omega_\varepsilon} \left(A(|\nabla v|)\nabla v - A(|\nabla u|)\nabla u\right) \cdot (\nabla v - \nabla u)\phi'(v - \varepsilon - u)dx \leq 0.$$

Observe now that

$$\left(A(|\nabla v|)\nabla v - A(|\nabla u|)\nabla u\right)\cdot(\nabla v - \nabla u)$$

$$= \left(A(|\nabla v|)|\nabla v| - A(|\nabla u|)|\nabla u|\right)(|\nabla v| - |\nabla u|)$$

$$+ \left(A(|\nabla v|) + A(|\nabla u|)\right)(|\nabla v||\nabla u| - \nabla v \cdot \nabla u) \geq 0.$$

By the last two estimates, we must have equality to zero throughout which yields $\nabla u = \nabla v$ in Ω_ε. This yields $u - v$ is constant in Ω_ε, which is impossible since $u - v = \varepsilon$ on $\partial\Omega_\varepsilon$ and $u < v - \varepsilon$ in Ω_ε. Hence, for all $\varepsilon > 0$, we have $u \geq v - \varepsilon$ in Ω. Letting $\varepsilon \to 0$ we reach the conclusion. □

Remark 2.7 N. Trudinger [Tru67] proved that if $\mathscr{A}$ is S-m-C, then $\mathscr{L}$ satisfies the weak Harnack inequality.

Definition 2.8 We say that $u \in W_{loc}^{1,1}(\Omega) \cap C(\Omega)$ is a positive solution of (2.1) if:

- $u > 0$, $\mathscr{A}(x, u, \nabla u) \in L^1_{loc}(\Omega)^N$, $\mathscr{L}u \in L^1_{loc}(\Omega)$, $(|x|^{-\alpha} * u^p)u^q \in L^1_{loc}(\Omega)$.
-

$$\int_\Omega \frac{u^p(y)}{1 + |y|^{-\alpha}} dy < \infty. \tag{2.5}$$

- for any $\phi \in C_c^\infty(\Omega)$, $\phi \geq 0$ we have

$$\int_\Omega \mathscr{A}(x, u, \nabla u) \cdot \nabla \phi \geq \int_\Omega (|x|^{-\alpha} * u^p)u^q \phi.$$

Condition (2.5) follows from the fact that $|x|^{-\alpha} * u^p < \infty$ in Ω.

2.2 A Priori Estimates

A key tool in our approach is the use of a priori estimates for solutions $u \in W^{1,1}_{loc}(\Omega) \cap C(\Omega)$ of the general inequality

$$- \operatorname{div}\left(\mathscr{A}(x, u, \nabla u)\right) \geq f(x) \quad \text{in } \Omega, \tag{2.6}$$

where $f \in L^1_{loc}(\Omega)$, $f \geq 0$. Solutions u of (2.6) are understood in the weak sense, that is,

$$\mathscr{A}(x, u, \nabla u) \in L^1_{loc}(\Omega)^N$$

and

$$\int_\Omega \mathscr{A}(x, u, \nabla u) \cdot \nabla \varphi \geq \int_\Omega f(x)\varphi \quad \text{for any } \varphi \in C_c^\infty(\Omega), \varphi \geq 0. \tag{2.7}$$

The main result in this section is stated below.

Proposition 2.9 *Let $\Omega \subset \mathbb{R}^N$ be an open set such that for some $R > 0$ we have*

$$B_{4R} \setminus B_{R/2} \subset \Omega \quad (\text{resp. } B_{2R} \subset \Omega).$$

Assume $\mathscr{A}$ is W-m-C and let $u \in W^{1,1}_{loc}(\Omega) \cap C(\Omega)$ be a positive solution of (2.6). Let $\phi \in C_c^\infty(\Omega)$ be a standard cutoff function such that:

- $0 \leq \phi \leq 1$ and $\operatorname{supp}\phi \subset B_{4R} \setminus B_{R/2}$ (resp. $\operatorname{supp}\phi \subset B_{2R}$);
- $\phi = 1$ in $B_{2R} \setminus B_R$ (resp. $\phi = 1$ in B_R);
- $|\nabla \phi| \leq \frac{C}{R}$ in Ω.

Then, for any $\lambda > m$, $0 \leq \theta \leq m - 1$ and $\ell > m - 1 - \theta$, there exists $C > 0$ independent of R such that

$$\int_{\Omega} f(x) u^{-\theta} \phi^{\lambda} \leq C R^{N-m-\frac{m-1-\theta}{\ell}N} \left(\int_{\Omega} u^{\ell} \phi^{\lambda} \right)^{\frac{m-1-\theta}{\ell}}. \tag{2.8}$$

In particular,

(i) If $B_{4R} \setminus B_{R/2} \subset \Omega$, then

$$\int_{B_{2R} \setminus B_R} f(x) u^{-\theta} \leq C R^{N-m-\frac{m-1-\theta}{\ell}N} \left(\int_{B_{4R} \setminus B_{R/2}} u^{\ell} \right)^{\frac{m-1-\theta}{\ell}}. \tag{2.9}$$

(ii) If $B_{2R} \subset \Omega$, then

$$\int_{B_{2R}} f(x) u^{-\theta} \leq C R^{N-m-\frac{m-1-\theta}{\ell}N} \left(\int_{B_{2R}} u^{\ell} \right)^{\frac{m-1-\theta}{\ell}}. \tag{2.10}$$

Proof Assume first $\theta > 0$ and let $\varphi = u^{-\theta} \phi^{\lambda}$ in (2.7). We find

$$\int_{\Omega} f(x) u^{-\theta} \phi^{\lambda} \leq \int_{\Omega} \mathscr{A}(x, u, \nabla u) \cdot \nabla \varphi$$

$$= -\theta \int_{\Omega} u^{-\theta-1} \phi^{\lambda} \mathscr{A}(x, u, \nabla u) \cdot \nabla u + \lambda \int_{\Omega} \phi^{\lambda-1} u^{-\theta} \mathscr{A}(x, u, \nabla u) \cdot \nabla \phi.$$

Using Definition 2.1 of a W-m-C operator, it follows that

$$\int_{\Omega} f(x) u^{-\theta} \phi^{\lambda} + C_1 \theta \int_{\Omega} u^{-\theta-1} \phi^{\lambda} |\mathscr{A}(x, u, \nabla u)|^{m'}$$

$$\leq \lambda \int_{\Omega} \phi^{\lambda-1} u^{-\theta} \mathscr{A}(x, u, \nabla u) \cdot \nabla \phi. \tag{2.11}$$

By Young's inequality , it follows that

$$\lambda \phi^{\lambda-1} u^{-\theta} \mathscr{A}(x, u, \nabla u) \cdot \nabla \phi \leq \frac{C_1 \theta}{2} u^{-\theta-1} \phi^{\lambda} |\mathscr{A}(x, u, \nabla u)|^{m'} + C(\theta, \lambda) u^{m-1-\theta} \phi^{\lambda-m} |\nabla \phi|^m.$$

Using the above inequality in (2.11), we find

$$\int_{\Omega} f(x) u^{-\theta} \phi^{\lambda} + \frac{C_1 \theta}{2} \int_{\Omega} u^{-\theta-1} \phi^{\lambda} |\mathscr{A}(x, u, \nabla u)|^{m'} \leq C \int_{\Omega} u^{m-1-\theta} \phi^{\lambda-m} |\nabla \phi|^m.$$

In particular, this yields

$$\int_\Omega f(x)u^{-\theta}\phi^\lambda + \frac{C_1\theta}{2}\int_\Omega u^{-\theta-1}\phi^\lambda|\mathscr{A}(x,u,\nabla u)|^{m'} \leq C\int_\Omega u^{m-1-\theta}\phi^{\lambda-m}|\nabla\phi|^m.$$

$$(2.12)$$

If $\theta = m - 1$, it follows from (2.12) and the properties of ϕ that

$$\int_\Omega f(x)u^{-\theta}\phi^\lambda \leq C\int_\Omega \phi^{\lambda-m}|\nabla\phi|^m \leq CR^{N-m},$$

which proves (2.8). Assume next $0 < \theta < m - 1$ and let

$$\gamma = \frac{\ell}{m-1-\theta} > 1. \tag{2.13}$$

From (2.12) and Hölder's inequality (in the following, γ' stands for Hölder's conjugate of γ), we find

$$\int_\Omega f(x)u^{-\theta}\phi^\lambda \leq C\int_\Omega u^{m-1-\theta}\phi^{\lambda-m}|\nabla\phi|^m$$

$$\leq C\Big(\int_{\mathrm{supp}\nabla\phi} u^\ell\phi^\lambda\Big)^{1/\gamma}\Big(\int_{\mathrm{supp}\nabla\phi}\phi^{\lambda-m\gamma'}|\nabla\phi|^{m\gamma'}\Big)^{1/\gamma'}$$

$$\leq CR^{\frac{N}{\gamma'}-m}\Big(\int_{\mathrm{supp}\nabla\phi} u^\ell\phi^\lambda\Big)^{1/\gamma}$$

$$= CR^{N-m-\frac{m-1-\theta}{\ell}N}\Big(\int_\Omega u^\ell\phi^\lambda\Big)^{\frac{m-1-\theta}{\ell}}.$$

It remains to discuss the case $\theta = 0$. We fix $\sigma \in (0, m - 1)$ such that

$$\tau = \frac{\ell}{(1+\sigma)(m-1)} > 1 \tag{2.14}$$

and let $\varphi = \phi^\lambda$ in (2.7). By Hölder's inequality, we have

$$\int_\Omega f(x)\phi^\lambda \leq \lambda\int_\Omega \phi^{\lambda-1}\mathscr{A}(x,u,\nabla u)\cdot\nabla\phi$$

$$\leq \lambda\Big(\int_\Omega u^{-\sigma-1}\phi^\lambda|\mathscr{A}(x,u,\nabla u)|^{m'}\Big)^{1/m'}\Big(\int_\Omega u^{(1+\sigma)(m-1)}\phi^{\lambda-m}|\nabla\phi|^m\Big)^{1/m}.$$

In particular, this yields

$$\int_\Omega f(x)\phi^\lambda \leq C\Big(\int_\Omega u^{-\sigma-1}\phi^\lambda|\mathscr{A}(x,u,\nabla u)|^{m'}\Big)^{1/m'}\Big(\int_\Omega u^{(1+\sigma)(m-1)}\phi^{\lambda-m}|\nabla\phi|^m\Big)^{1/m}.$$

$$(2.15)$$

Let now $\varphi = u^{-\sigma}\phi^\lambda$ in (2.7). Using the same argument as above, we arrive at (2.12) in which θ is replaced now with σ. In particular, we find

$$\int_\Omega u^{-\sigma-1}\phi^\lambda |\mathscr{A}(x,u,\nabla u)|^{m'} \le C\int_\Omega u^{m-1-\sigma}\phi^{\lambda-m}|\nabla\phi|^m. \tag{2.16}$$

Combining (2.15) and (2.16), we deduce

$$\int_\Omega f(x)\phi^\lambda \le C\Big(\int_\Omega u^{m-1-\sigma}\phi^{\lambda-m}|\nabla\phi|^m\Big)^{1/m'}\Big(\int_\Omega u^{(1+\sigma)(m-1)}\phi^{\lambda-m}|\nabla\phi|^m\Big)^{1/m}. \tag{2.17}$$

Using Hölder's inequality with exponents γ and τ defined in (2.13) and (2.14) (in which again we replace θ with σ), we obtain

$$\int_\Omega u^{m-1-\sigma}\phi^{\lambda-m}|\nabla\phi|^m \le \Big(\int_\Omega u^\ell\phi^\lambda\Big)^{1/\gamma}\Big(\int_\Omega \phi^{\lambda-m\gamma'}|\nabla\phi|^{m\gamma'}\Big)^{1/\gamma'} \tag{2.18}$$

$$\le CR^{\frac{N}{\gamma'}-m}\Big(\int_\Omega u^\ell\phi^\lambda\Big)^{1/\gamma},$$

and

$$\int_\Omega u^{(1+\sigma)(m-1)}\phi^{\lambda-m}|\nabla\phi|^m \le \Big(\int_\Omega u^\ell\phi^\lambda\Big)^{1/\tau}\Big(\int_\Omega \phi^{\lambda-m\tau'}|\nabla\phi|^{m\tau'}\Big)^{1/\tau'} \tag{2.19}$$

$$\le CR^{\frac{N}{\tau'}-m}\Big(\int_\Omega u^\ell\phi^\lambda\Big)^{1/\tau}.$$

Now, (2.8) follows by combining (2.17)–(2.19). □

Letting $\theta = m - 1$ in Proposition 2.9, we derive the following a priori estimate.

Lemma 2.10 *Assume $\mathscr{A}$ is W-m-C for some $m > 1$ and let u be a positive solution of (2.1) in $\mathbb{R}^N \setminus \overline{B}_1$. Then*

$$\Big(\int_{B_{2R}\setminus B_1} u^p\Big)\Big(\int_{B_{2R}\setminus B_R} u^{q-m+1}\Big) \le CR^{N-m+\alpha} \quad \text{for all } R > 2.$$

Proof Taking $\theta = m - 1$ in Proposition 2.9, we find

$$\int_{B_{2R}\setminus B_R} (|x|^{-\alpha} * u^p)u^{q-m+1} \le CR^{N-m} \quad \text{for all } R > 2. \tag{2.20}$$

If $x \in B_{2R}\setminus B_R$ and $y \in B_{2R}\setminus B_1$, then $|x-y| \le |x|+|y| \le 4R$, so

$$(|x|^{-\alpha} * u^p)(x) \geq C \int_{B_{2R} \setminus B_1} \frac{u^p(y)}{|x-y|^\alpha} dy$$

$$\geq C \int_{B_{2R} \setminus B_1} \frac{u^p(y)}{(4R)^\alpha} dy \qquad (2.21)$$

$$\geq C R^{-\alpha} \int_{B_{2R} \setminus B_1} u^p(y) dy.$$

The proof concludes by combining (2.21) and (2.20). □

Proposition 2.11 *Suppose $N > m$, $\mathscr{A}$ is (H_m) and let $u \in C(\mathbb{R}^N \setminus \overline{B}_1) \cap W^{1,1}_{loc}(\mathbb{R}^N \setminus \overline{B}_1)$ be a positive solution of*

$$\mathscr{L}u \geq 0 \quad in \ \mathbb{R}^N \setminus \overline{B}_1.$$

Then, there exists $c > 0$ such that

$$u(x) \geq c|x|^{-\frac{N-m}{m-1}} \quad in \ \mathbb{R}^N \setminus B_2.$$

Proof Let $c > 0$ be such that $u \geq c$ on ∂B_2. Take $\delta > 0$. Then, for $k > 2$ large, we have $\delta > c|x|^{-(N-m)/(m-1)}$ on ∂B_k. Then, letting $v(x) = c|x|^{-(N-m)/(m-1)}$ we have $u + \delta > v$ on $\partial(B_k \setminus \overline{B}_2)$. We apply Proposition 2.6 to $u + \delta$ and v on $B_k \setminus \overline{B}_2$ to derive $u + \delta \geq v$ on $B_k \setminus \overline{B}_2$. We next let $\delta \to 0$ and by the fact that $k > 2$ can be arbitrarily large, we deduce the conclusion. □

Lemma 2.12 *Let $\alpha \in (0, N)$ and $\rho > 0$.*

(i) *If $f \in L^1_{loc}(\mathbb{R}^N \setminus \overline{B}_\rho)$, $f \geq 0$, then there exists $C > 0$ such that*

$$\int_{\mathbb{R}^N \setminus B_\rho} \frac{f(y)}{|x-y|^\alpha} dy \geq C|x|^{-\alpha} \quad for \ any \ x \in \mathbb{R}^N \setminus B_{2\rho}.$$

(ii) *If $f(x) \geq c|x|^{-\beta}$ in $\mathbb{R}^N \setminus B_{2\rho}$ for some $c, \beta > 0$, then*

$$\begin{cases} |x|^{-\alpha} * f = \infty & if \ \beta \leq N - \alpha \\ |x|^{-\alpha} * f \geq C|x|^{N-\alpha-\beta} & if \ \beta > N - \alpha \end{cases} \quad in \ \mathbb{R}^N \setminus B_{2\rho},$$

for some constant $C > 0$.

(iii) *If $f \in L^1(\mathbb{R}^N) \cap C(\mathbb{R}^N)$ and $f(x) \leq c|x|^{-\beta}$ in $\mathbb{R}^N \setminus B_\rho$ for some $c > 0$ and $\beta > N - \alpha > 0$, then for some constant $C > 0$, we have*

$$\int_{\mathbb{R}^N} \frac{f(y)}{|x-y|^\alpha} dy \leq \begin{cases} C|x|^{N-\alpha-\beta} & \text{if } N-\alpha < \beta < N \\ C|x|^{-\alpha} \log|x| & \text{if } \beta = N \\ C|x|^{-\alpha} & \text{if } \beta > N \end{cases} \quad \text{in } \mathbb{R}^N \setminus B_{2\rho}.$$

Proof

(i) For any $x \in \mathbb{R}^N \setminus B_{2\rho}$, we have

$$|x|^{-\alpha} * f \geq \int_{3\rho/2<|y|<2\rho} \frac{f(y)}{|x-y|^\alpha} dy \geq C \int_{3\rho/2<|y|<2\rho} \frac{f(y)}{|2x|^\alpha} dy = C|x|^{-\alpha}.$$

(ii) Since $|x-y| \leq |x| + |y| \leq 3|y|$ if $|y| \geq 2|x|$, we have

$$|x|^{-\alpha} * f \geq c \int_{|y|\geq 2|x|} \frac{|y|^{-\beta}}{|x-y|^\alpha} dy \geq C \int_{|y|\geq 2|x|} |y|^{-\alpha-\beta} dy = C \int_{2|x|}^\infty t^{N-\alpha-\beta} \frac{dt}{t},$$

and the conclusion follows.

(iii) Let $|x| \geq 2\rho$. We have

$$\int_{\mathbb{R}^N} \frac{f(y)}{|x-y|^\alpha} dy = \left\{ \int_{|y|\geq 2|x|} + \int_{\frac{1}{2}|x|\leq|y|\leq 2|x|} + \int_{|y|\leq|x|/2} \right\} \frac{f(y)}{|x-y|^\alpha} dy.$$

For $|y| \geq 2|x|$ we have $|x-y| \geq |y| - |x| \geq |y|/2$ so that

$$\int_{|y|\geq 2|x|} \frac{f(y)}{|x-y|^\alpha} dy \leq C \int_{|y|\geq 2|x|} \frac{dy}{|y|^{\alpha+\beta}} \leq C|x|^{N-\alpha-\beta}.$$

Similarly we estimate

$$\int_{\frac{1}{2}|x|\leq|y|\leq 2|x|} \frac{f(y)}{|x-y|^\alpha} dy \leq c|x|^{-\beta} \int_{\frac{1}{2}|x|\leq|y|\leq 2|x|} \frac{dy}{|x-y|^\alpha}$$

$$\leq c|x|^{-\beta} \int_{|y-x|\leq 3|x|} \frac{dy}{|x-y|^\alpha} = C|x|^{N-\alpha-\beta}.$$

Finally, if $|y| \leq |x|/2$ then $|x-y| \geq |x| - |y| \geq |x|/2$. Hence

$$\int_{|y|\leq|x|/2} \frac{f(y)}{|x-y|^\alpha} dy \leq C|x|^{-\alpha} \int_{|y|\leq|x|/2} f(y) dy$$

$$\leq C|x|^{-\alpha} \left\{ \int_{|y|\leq 1} f(y) dy + \int_{1<|y|\leq|x|/2} f(y) dy \right\}$$

$$\leq C|x|^{-\alpha}\left\{1 + C\int_{1<|y|\leq|x|/2}|y|^{-\beta}dy\right\}$$

$$\leq C|x|^{-\alpha} + C\begin{cases}|x|^{N-\alpha-\beta} & \text{if } \beta < N, \\ |x|^{-\alpha}\log|x| & \text{if } \beta = N, \\ |x|^{-\alpha} & \text{if } \beta > N.\end{cases}$$

The result now follows by combining the above three estimates.   □

2.3 Nonexistence Results

Our first nonexistence result concerns the general case where $\mathscr{A}$ is W-m-C.

Theorem 2.13 *Let $\Omega = \mathbb{R}^N \setminus \overline{B}_1$. Assume $\mathscr{A}$ is W-m-C, $N > m > 1$ and one of the following condition holds:*

(i) $p + q > m - 1$, $q \leq m - 1$ and $q < m - 1 - \frac{\alpha-m}{N}p$.
(ii) $p + q > m - 1$, $q < m - 1$ and $q = m - 1 - \frac{\alpha-m}{N}p$.
(iii) $p + q \leq m - 1$.

Then (2.1) has no positive solutions.

Proof

(i) Assume first that $q = m - 1$ which also implies $\alpha < m$. Then by Lemma 2.10, we deduce

$$\int_{B_{2R}\setminus B_1} u^p dx \leq C R^{\alpha-m} \quad \text{for all } R > 2.$$

Letting $R \to \infty$ in the above estimate and using the fact that $\alpha < m$, it follows that (2.1) has no positive solutions.

Assume next that $q < m - 1$ and by Hölder's inequality, we estimate

$$\int_{B_{2R}\setminus B_R} 1 \leq \left(\int_{B_{2R}\setminus B_R} u^p\right)^{\frac{m-1-q}{p+m-1-q}}\left(\int_{B_{2R}\setminus B_R} u^{q-m+1}\right)^{\frac{p}{p+m-1-q}}, \quad (2.22)$$

which we rewrite as

$$C R^N \leq \left[\left(\int_{B_{2R}\setminus B_R} u^p\right)\left(\int_{B_{2R}\setminus B_R} u^{q-m+1}\right)\right]^{\frac{m-1-q}{p+m-1-q}}\left(\int_{B_{2R}\setminus B_R} u^{q-m+1}\right)^{\frac{p-m+1+q}{p+m-1-q}}.$$

Now, by Lemma 2.10, we deduce

$$CR^N \leq C\big(R^{N-m+\alpha}\big)^{\frac{m-1-q}{p+m-1-q}} \Big(\int_{B_{2R}\setminus B_R} u^{q-m+1}\Big)^{\frac{p-m+1+q}{p+m-1-q}},$$

which yields

$$\int_{B_{2R}\setminus B_R} u^{q-m+1} \geq C R^{N+\frac{(N+m-\alpha)(m-1-q)}{p-m+1+q}} \qquad \text{for } R > 2. \tag{2.23}$$

Again by Lemma 2.10, we have

$$\Big(\int_{B_{2R}\setminus B_1} u^p\Big)\Big(\int_{B_{2R}\setminus B_R} u^{q-m+1}\Big) \leq C R^{N-m+\alpha} \qquad \text{for all } R > 2. \tag{2.24}$$

Therefore,

$$\int_{B_{2R}\setminus B_R} u^{q-m+1} dx \leq \frac{C R^{N-m+\alpha}}{\int_{B_{2R}\setminus B_1} u^p} \leq \frac{C R^{N-m+\alpha}}{\int_{B_4\setminus B_1} u^p} \leq C R^{N-m+\alpha}. \tag{2.25}$$

From (2.23) and (2.25), we deduce

$$C_1 R^{N+\frac{(N+m-\alpha)(m-1-q)}{p-m+1+q}} \leq \int_{B_{2R}\setminus B_R} u^{q-m+1} \leq C_2 R^{N-m+\alpha} \qquad \text{for all } R > 2.$$

Since $q < m - 1 - \frac{\alpha-m}{N} p$, the above inequality cannot hold for large $R > 2$. Hence, (2.1) cannot have positive solutions.

(ii) Assume $0 < q = m - 1 - \frac{\alpha-m}{N} p$. With a similar approach as above, we have

$$C_1 R^{N-m+\alpha} \leq \int_{B_{2R}\setminus B_R} u^{q-m+1} \leq C_2 R^{N-m+\alpha} \qquad \text{for all } R > 2. \tag{2.26}$$

Since $\frac{p}{p+m-1-q} = \frac{N}{N-m+\alpha}$, estimate (2.22) yields

$$CR^N \leq \Big(\int_{B_{2R}\setminus B_R} u^p\Big)^{\frac{m-1-q}{p+m-1-q}} \cdot CR^N \qquad \text{for all } R > 2.$$

Hence

$$\int_{B_{2R}\setminus B_R} u^p \geq C \qquad \text{for all } R > 2,$$

which shows that $\int_{\mathbb{R}^N\setminus B_1} u^p = \infty$ and then $\int_{B_{2R}\setminus B_1} u^p \to \infty$ as $r \to \infty$. Further, from (2.25) we have

$$\int_{B_{2R}\setminus B_R} u^{q-m+1} \leq \frac{C R^{N-m+\alpha}}{\int_{B_{2R}\setminus B_1} u^p} = o(R^{N-m+\alpha}) \qquad \text{as } R \to \infty,$$

which contradicts the first estimate in (2.26).

(iii) Assume first that $p + q = m - 1$. By Lemma 2.10, we have

$$\left(\int_{B_{2R}\setminus B_1} u^p\right)\left(\int_{B_{2R}\setminus B_R} u^{-p}\right) \leq C R^{N-m+\alpha} \quad \text{for all } R > 2. \tag{2.27}$$

On the other hand, by Hölder's inequality, we deduce

$$C R^{2N} = \left(\int_{B_{2R}\setminus B_R} 1\right)^2 \leq \left(\int_{B_{2R}\setminus B_R} u^p\right)\left(\int_{B_{2R}\setminus B_R} u^{-p}\right) \quad \text{for all } R > 2. \tag{2.28}$$

Now, (2.27) and (2.28) cannot hold for $R > 2$ sufficiently large. This shows that (2.1) cannot have positive solutions.

Assume now that $p + q < m - 1$. We apply Hölder's inequality to derive (2.22) which we may rewrite as

$$C R^N \leq \left(\int_{B_{2R}\setminus B_R} u^p\right)^{\frac{m-1-p-q}{p+m-1-q}} \left[\left(\int_{B_{2R}\setminus B_R} u^p\right)\left(\int_{B_{2R}\setminus B_R} u^{q-m+1}\right)\right]^{\frac{p}{p+m-1-q}}.$$

Using the estimate in Lemma 2.10, we find

$$R^N \leq C \left(R^{N-m+\alpha}\right)^{\frac{p}{p+m-1-q}} \left(\int_{B_{2R}\setminus B_R} u^p\right)^{\frac{m-1-p-q}{p+m-1-q}} \quad \text{for all } R > 2,$$

which implies

$$\int_{B_{2R}\setminus B_R} u^p \geq C R^{N+\frac{p(N+m-\alpha)}{m-1-p-q}} \quad \text{for all } R > 2. \tag{2.29}$$

On the other hand, Eq. (2.5) yields

$$\int_{|x|>1} \frac{u^p(x)}{|x|^\alpha} dx < \infty,$$

which further implies

$$\int_{B_{2R}\setminus B_R} u^p \leq C R^\alpha \quad \text{for all } R > 2.$$

Combining this with (2.29), we find

$$C_1 R^\alpha \geq \int_{B_{2R}\setminus B_R} u^p \geq C_2 R^{N+\frac{p(N+m-\alpha)}{m-1-p-q}} \quad \text{for all } R > 2,$$

which gives a contradiction as $\alpha < N + \frac{p(N+m-\alpha)}{m-1-p-q}$. $\square$

In case $\mathscr{A}$ is (H_m), we may obtain further nonexistence results.

Theorem 2.14 *Let* $\Omega = \mathbb{R}^N \setminus \overline{B}_1$. *Assume* $\mathscr{A}$ *is* (H_m) *for some* $m > 1$.

(i) *If* $m \geq N$, *then* (2.1) *has no positive solutions.*
(ii) *If* $N > m > 1$ *and one of the following conditions hold:*

    (ii1) $0 < p \leq \frac{(N-\alpha)(m-1)}{N-m}$;
    (ii2) $m - 1 < q \leq \frac{(N-\alpha)(m-1)}{N-m}$ *and* $\alpha < m$;
    (ii3) $m - 1 \leq p + q \leq \frac{(2N-\alpha)(m-1)}{N-m}$;

then (2.1) *has no positive solutions.*

Proof Assume by contradiction that u is a positive solution of (2.1).

(i) By Proposition 2.11, we have $u \geq c$ in $\mathbb{R}^N \setminus B_2$, for some constant $c > 0$. Therefore,

$$\int_{|y|>2} \frac{u^p(y)}{1+|y|^\alpha} dy \geq c \int_{|y|>2} \frac{dy}{2|y|^\alpha} = \infty,$$

which contradicts (2.5).

(ii1) By Proposition 2.11, we deduce $u \geq c|x|^{-\frac{N-m}{m-1}}$ in $\mathbb{R}^N \setminus B_2$, for some constant $c > 0$. Since $p \leq \frac{(N-\alpha)(m-1)}{N-m}$, by Lemma 2.12(ii), we have $|x|^{-\alpha} * u^p = \infty$ for all $x \in \mathbb{R}^N \setminus B_2$, which contradicts the fact that $(|x|^{-\alpha} * u^p)u^q \in L^1_{loc}(\Omega)$.
(ii2) Assume $m-1 < q \leq \frac{(N-\alpha)(m-1)}{N-m}$ which further yields $\alpha < m$. By Lemma 2.12(i), we have

$$|x|^{-\alpha} * u^p \geq C|x|^{-\alpha} \quad \text{in } \mathbb{R}^N \setminus B_2.$$

Thus, u satisfies

$$\mathscr{L}u \geq C|x|^{-\alpha}u^q \quad \text{in } \mathbb{R}^N \setminus B_2. \tag{2.30}$$

We now apply Proposition 2.9 with $f = |x|^{-\alpha}u^q$, $\theta = 0$ and $\ell = q > m - 1$. We have

$$\int |x|^{-\alpha}u^q\phi^\lambda \leq C R^{N-m-\frac{m-1}{q}N}\left(\int u^q\phi^\lambda\right)^{\frac{m-1}{q}}$$

$$\leq C R^{N-m-\frac{(N-\alpha)(m-1)}{q}}\left(\int |x|^{-\alpha}u^q\phi^\lambda\right)^{\frac{m-1}{q}}.$$

Since $\phi = 1$ on $B_{2R} \setminus B_R$, using the estimate $u \geq C|x|^{-\frac{N-m}{m-1}}$ in $\mathbb{R}^N \setminus B_2$, it follows that

$$\left(\int |x|^{-\alpha} u^q \phi^\lambda \right)^{\frac{m-1}{q}} \leq C R^{N-m-\frac{(N-\alpha)(m-1)}{q}}. \tag{2.31}$$

If $q < (N-\alpha)(m-1)/(N-m)$, the right-hand side of (2.31) tends to zero as $R \to \infty$ which implies

$$\lim_{R \to \infty} \int |x|^{-\alpha} u^q \phi^\lambda = 0, \quad \text{that is,} \quad \int |x|^{-\alpha} u^q = 0$$

and thus $u \equiv 0$. If $q = \alpha(m-1)/(N-m)$, then (2.30) yields

$$\mathcal{L} u \geq C|x|^{-N} \quad \text{in } \mathbb{R}^N \setminus B_2. \tag{2.32}$$

We claim that for some $c > 0$, one has

$$u(x) \geq c|x|^{-\frac{N-m}{m-1}} \left(\ln|x| \right)^{\frac{1}{m-1}} \quad \text{in } \mathbb{R}^N \setminus B_2. \tag{2.33}$$

Indeed, let $s \in (0, 1)$ be small enough such that $s < \min_{|x|=2} u(x)$ and for $n \geq 3$ let u_n be a radial function that satisfies

$$\begin{cases} \mathcal{L} u_n = C|x|^{-N} & \text{in } B_n \setminus \overline{B}_2, \\ u_n = s & \text{on } \partial B_2, \\ u_n = 0 & \text{on } \partial B_n. \end{cases}$$

Then $u_n \leq u_{n+1} \leq u$ in $B_n \setminus B_2$ and $\{u_n\}$ is uniformly bounded in any compact subset of $\mathbb{R}^N \setminus \overline{B}_2$. Thus, $\{u_n\}$ converges in $C^1_{loc}(B_R \setminus \overline{B}_2)$ to some $v \in C^1(\mathbb{R}^N \setminus \overline{B}_2)$ that satisfies $u \geq v$ in $\mathbb{R}^N \setminus \overline{B}_2$ and

$$\begin{cases} \mathcal{L} v = C|x|^{-N} & \text{in } \mathbb{R}^N \setminus \overline{B}_2, \\ v = s & \text{on } \partial B_2. \end{cases}$$

Thus,

$$-A(|v'(r)|)v'(r) = Cr^{1-N} \ln r + cr^{1-N} \quad \text{for all } r > 2,$$

where $C > 0$. From the above equality, it follows that $v'(r) < 0$ for $r > 2$ large, so there exists $\ell = \lim_{r \to \infty} v(r) \geq 0$. Further, there exists $\rho > 2$ and $C_0 > 0$ such that

$$|v'(r)|^{m-1} \geq C_0 r^{1-N} \ln r \quad \text{for all } r \geq \rho.$$

This yields

$$-v'(r) \geq C_1 r^{\frac{1-N}{m-1}} (\ln r)^{\frac{1}{m-1}} \quad \text{for all } r \geq \rho.$$

We integrate the above estimate over $[r, R]$ and then let $R \to \infty$. Using the fact that v is continuous and positive, we obtain

$$v(r) \geq C_2 r^{-\frac{N-m}{m-1}} (\ln r)^{\frac{1}{m-1}} \quad \text{for all } r \geq 2.$$

Since $u \geq v$ in $\mathbb{R}^N \setminus \overline{B}_2$, the claim follows.

Finally, using the estimate (2.33) back into (2.31), it follows that (since $\phi = 1$ on $B_{2R} \setminus B_R$)

$$C(\ln R)^{\frac{m-\alpha}{N-\alpha}} \leq \left(\int_{B_{2R} \setminus B_R} |x|^{-\alpha} u^q \right)^{1 - \frac{m-1}{q}} \leq C.$$

which is a contradiction.

(ii3) Assume first that $p+q < \frac{(N+\alpha)(m-1)}{N-m}$. By Hölder's inequality, we estimate

$$\left(\int_{B_{2R} \setminus B_R} u^p \right) \left(\int_{B_{2R} \setminus B_R} u^{q-m+1} \right) \geq \left(\int_{B_{2R} \setminus B_R} u^{\frac{p+q-m+1}{2}} \right)^2 \quad \text{for all } R > 2.$$

Using Lemma 2.10 together with the estimate $u \geq C|x|^{-\frac{N-m}{m-1}}$ in $\mathbb{R}^N \setminus B_2$, we find

$$C_1 R^{N-m+\alpha} \geq \left(\int_{B_{2R} \setminus B_R} u^{\frac{p+q-m+1}{2}} \right)^2 \geq C_2 R^{2N - \frac{(p+q-m+1)(N-m)}{m-1}} \quad \text{for all } R > 2.$$

However, this is a contradiction as $N - m + \alpha < 2N - \frac{(p+q-m+1)(N-m)}{m-1}$.

Assume now that $p + q = \frac{(2N-\alpha)(m-1)}{N-m}$. If $q < 0$, the nonexistence of a positive solution follows from Theorem 2.13(i). Hence, it remains to discuss the case $q \geq 0$. By Proposition 2.11, we deduce $u \geq c|x|^{-\frac{N-m}{m-1}}$ in $\mathbb{R}^N \setminus B_2$, for some constant $c > 0$. If $p\frac{N-m}{m-1} \leq N - \alpha$, by Lemma 2.12(ii), we find $|x|^{-\alpha} * u^p = \infty$ in $\mathbb{R}^N \setminus B_1$, contradiction. Assume in the following that $p\frac{N-m}{m-1} > N - \alpha$. By the last estimate in Lemma 2.12(ii), it follows that

$$|x|^{-\alpha} * u^p \geq C|x|^{N-\alpha - p\frac{N-m}{m-1}} \quad \text{in } \mathbb{R}^N \setminus B_2.$$

Thus, u satisfies

$$\mathscr{L}u \geq (|x|^{-\alpha} * u^p)u^q \geq C|x|^{N-\alpha - (p+q)\frac{N-m}{m-1}} = C|x|^{-N} \quad \text{in } \mathbb{R}^N \setminus B_2.$$

Hence, u satisfies (2.32) and then (2.33) holds.

Now, by (2.33), Lemma 2.10 and Hölder's inequality, for $R > 2$ we have

$$
CR^{N-m+\alpha} \geq \left(\int_{B_{2R} \setminus B_R} u^p \right) \left(\int_{B_{2R} \setminus B_R} u^{q-m+1} \right)
$$

$$
\geq \left(\int_{B_{2R} \setminus B_R} u^{\frac{p+q-m+1}{2}} \right)^2
$$

$$
\geq CR^{N-m+\alpha} (\ln R)^{\frac{N+m-\alpha}{N-m}},
$$

which is a contradiction for $R > 2$ large. $\square$

2.4   Existence of Solutions in $\mathbb{R}^N$

In the following, we assume that $\mathscr{A}$ has the form

$$
\mathscr{A}(x, u, \nabla u) = A(|\nabla u|) \nabla u \tag{2.34}
$$

for some function $A \in C[0, \infty) \cap C^1(0, \infty)$ that satisfies

$$
A(t) \geq Ct^{m-2} \text{ for small enough } t > 0, \tag{2.35}
$$

and

$$
\frac{tA'(t)}{A(t)} \leq m - 2 \quad \text{for small enough } t > 0, \tag{2.36}
$$

where $C > 0$ and $m > 1$.

Clearly, for any $m > 1$, the m-Laplace operator and the m-mean curvature operator as given in (2.3) satisfy (2.34)–(2.36). More generally, if $A(t) = t^{m-2} f(t)$, where f is continuously differentiable at $t = 0$ and such that $f(0) > 0 \geq f'(t)$ for small enough $t > 0$, then (2.35)–(2.36) are fulfilled. Indeed, one can easily check that

$$
\frac{tA'(t)}{A(t)} = m - 2 + \frac{tf'(t)}{f(t)} \leq m - 2 \quad \text{for small enough } t > 0.
$$

Theorem 2.15 *Assume (2.34)–(2.36), $N > m > 1$ and one of the following conditions hold:*

(i) $p > \frac{N(m-1)}{N-m}$, $q = m - 1$ and $\alpha = m$.

(ii) $p \geq \frac{N(m-1)}{N-m}$, $q > m - 1 - p\frac{\alpha - m}{N}$.

(iii) $p > \frac{(N-\alpha)(m-1)}{N-m}$, $q > \frac{(N-\alpha)(m-1)}{N-m}$ and $p + q > \frac{(2N-\alpha)(m-1)}{N-m}$.

Then, (2.1) has a bounded radial solution in $\Omega = \mathbb{R}^N$.

Proof If $u = u(|x|) = u(r)$ is a radial function, one has

$$-\mathscr{L}u = \text{div}(A(|\nabla u|)\nabla u)$$

$$= u''(r) \cdot \left[A'(|u'(r)|) \cdot |u'(r)| + A(|u'(r)|) \right] + \frac{N-1}{r} \cdot A(|u'(r)|) \cdot u'(r).$$
$$(2.37)$$

We look for radial solutions of the form $u(r) = \varepsilon(1+r)^{-\gamma}$ for some $\varepsilon, \gamma > 0$ to be chosen later. Note that $u(r)$ is decreasing and let $t = |u'(r)|$ for convenience. According to (2.37), we have

$$\text{div}(A(|\nabla u|)\nabla u) = u''(r) \cdot \left[tA'(t) + A(t) \right] - \frac{N-1}{r} \cdot tA(t)$$

$$= \frac{tA(t)}{r} \cdot \left[-\frac{ru''(r)}{u'(r)} \cdot \frac{tA'(t) + A(t)}{A(t)} + 1 - N \right]$$

$$= \frac{tA(t)}{r} \cdot \left[\frac{(\gamma+1)r}{r+1} \cdot \frac{tA'(t) + A(t)}{A(t)} + 1 - N \right].$$

Since $t = \varepsilon\gamma(1+r)^{-\gamma-1} \le \varepsilon\gamma$, one has that $t \to 0$ uniformly as $\varepsilon \to 0$. For ε sufficiently small, our assumption (2.36) becomes applicable and

$$\text{div}(A(|\nabla u|)\nabla u) \le \frac{tA(t)}{r} \cdot \left[(\gamma+1)(m-1) + 1 - N \right]$$

$$= \frac{tA(t)}{r} \cdot \left[\gamma(m-1) + m - N \right].$$

As long as $0 < \gamma < \frac{N-m}{m-1}$, the constant in the square brackets is negative, so we get

$$\mathscr{L}u = -\text{div}(A(|\nabla u|)\nabla u) \ge \frac{C_0 tA(t)}{r} \ge \frac{C_1 t^{m-1}}{1+r}$$
$$(2.38)$$

$$= C_2 \varepsilon^{m-1}(1+r)^{-\gamma(m-1)-m}$$

for any $0 < \gamma < \frac{N-m}{m-1}$ and all sufficiently small $\varepsilon > 0$.

(i) Let us choose the parameter $\gamma > 0$ so that

$$\frac{N}{p} < \gamma < \frac{N-m}{m-1}.$$
$$(2.39)$$

Since $u(r) = \varepsilon(1+r)^{-\gamma}$ with $\gamma p > N > N - \alpha$, it follows by Lemma 2.12(iii) that

$$(|x|^{-\alpha} * u^p) \cdot u^q \le C_3 \varepsilon^p (1+r)^{-\alpha} \cdot u^q = C_4 \varepsilon^{p+q} (1+r)^{-\alpha-\gamma q}. \qquad (2.40)$$

Using our estimate (2.38) and the fact that $q = m - 1$, we also have

$$\mathscr{L}u \ge C_2 \varepsilon^{m-1} (1+r)^{-\gamma(m-1)-m} = C_2 \varepsilon^{m-1} (1+r)^{-\gamma q - m}.$$

Since $\alpha = m$ and $p + q > m - 1$, one may then combine the last two estimates to get

$$\mathscr{L}u \ge C_2 \varepsilon^{m-1} (1+r)^{-\gamma q - m} \ge C_4 \varepsilon^{p+q} (1+r)^{-\gamma q - \alpha} \ge (|x|^{-\alpha} * u^p) \cdot u^q,$$

for all sufficiently small $\varepsilon > 0$. This completes the proof of part (i).

(ii) We choose the parameter $\gamma > 0$ so that

$$\max\left\{ \frac{N+m-\alpha}{p+q-m+1}, \frac{N-\alpha}{p} \right\} < \gamma < \frac{N}{p} \le \frac{N-m}{m-1}. \qquad (2.41)$$

Since $\gamma < \frac{N-m}{m-1}$, our estimate (2.38) is still valid. Since $N - \alpha < \gamma p < N$, however, one has

$$(|x|^{-\alpha} * u^p) \cdot u^q \le C_3 \varepsilon^p (1+r)^{N-\alpha-\gamma p} \cdot u^q = C_4 \varepsilon^{p+q} (1+r)^{N-\alpha-\gamma(p+q)}$$
$$(2.42)$$

by Lemma 2.12(iii). Using our estimate (2.38) along with (2.41), we conclude that

$$\mathscr{L}u \ge C_2 \varepsilon^{m-1} (1+r)^{-\gamma(m-1)-m} \ge C_4 \varepsilon^{p+q} (1+r)^{N-\alpha-\gamma(p+q)} \ge (|x|^{-\alpha} * u^p) \cdot u^q$$
$$(2.43)$$

for all small enough $\varepsilon > 0$, since $p + q > m - 1$. This completes the proof of part (ii).

(iii) We proceed as above, this time we look for a solution of the form

$$u(r) = \varepsilon (1+r)^{-\gamma} \cdot \left(1 - \frac{k}{1 + \log(1+r)} \right), \qquad (2.44)$$

where $\gamma = \frac{N-m}{m-1} > 0$ and $\varepsilon, k > 0$ are sufficiently small. As one can easily check,

$$u'(r) = -\frac{\gamma u(r)}{1+r} + \frac{\varepsilon k (1+r)^{-\gamma-1}}{(1+\log(1+r))^2} \qquad (2.45)$$

$$= -\varepsilon \cdot \frac{\gamma (\log(1+r))^2 + \gamma(2-k)\log(1+r) + [\gamma - k(\gamma+1)]}{(1+r)^{\gamma+1} \cdot (1+\log(1+r))^2}.$$

If we take $0 < k < \frac{\gamma}{\gamma+1}$, then the numerator is the sum of nonnegative terms, so $u(r)$ is decreasing and we also have the estimate

$$\frac{C_0 \varepsilon}{(1+r)^{\gamma+1} \cdot (1 + \log(1+r))^2} \leq |u'(r)| \leq \frac{C_1 \varepsilon}{(1+r)^{\gamma+1}} \tag{2.46}$$

for some fixed constants $C_0, C_1 > 0$. Next, we differentiate (2.45) to find that

$$u''(r) + \frac{\gamma u'(r)}{1+r} = \frac{\gamma u(r)}{(1+r)^2} - \frac{\varepsilon k(\gamma+1)(1+r)^{-\gamma-2}}{(1+\log(1+r))^2} - \frac{2\varepsilon k(1+r)^{-\gamma-2}}{(1+\log(1+r))^3}. \tag{2.47}$$

Letting $z = \log(1+r)$ for convenience, one may express the last equation in the form

$$u''(r) + \frac{\gamma u'(r)}{1+r} = \frac{\gamma \varepsilon(1-k+z)}{(1+r)^{\gamma+2}(1+z)} - \frac{\varepsilon k(\gamma+1)}{(1+r)^{\gamma+2}(1+z)^2} - \frac{2\varepsilon k}{(1+r)^{\gamma+2}(1+z)^3}$$

$$= \varepsilon \cdot \frac{\gamma z^3 + \gamma(3-k)z^2 + (3\gamma - 3\gamma k - k)z + (\gamma - 2\gamma k - 3k)}{(1+r)^{\gamma+2}(1+z)^3}.$$

If we take $0 < k < \frac{\gamma}{2\gamma+3} < \frac{3\gamma}{3\gamma+1}$, then the numerator is the sum of nonnegative terms. Assuming that $k > 0$ is sufficiently small, we thus have

$$u''(r) \geq -\frac{\gamma u'(r)}{1+r} \geq 0. \tag{2.48}$$

On the other hand, one may combine equations (2.44), (2.45) and (2.47) to find that

$$u''(r) + (\gamma+1) \cdot \frac{u'(r)}{1+r} = -\frac{\varepsilon k(\gamma \log(1+r) + \gamma + 2)}{(1+r)^{\gamma+2} \cdot (1+\log(1+r))^3} \leq 0. \tag{2.49}$$

Next, we employ our identity (2.37) which gives

$$\mathrm{div}(A(|\nabla u|)\nabla u) = u''(r) \cdot \left[t A'(t) + A(t) \right] + \frac{N-1}{r} \cdot A(t) \cdot u'(r)$$

$$= A(t) \cdot \left[u''(r) \cdot \frac{t A'(t) + A(t)}{A(t)} + \frac{N-1}{r} \cdot u'(r) \right]$$

with $t = |u'(r)|$. If we let ε approach zero, then t converges to zero uniformly by (2.46) and our assumption (2.36) becomes applicable. Since $u''(r) \geq 0$ and $u'(r) \leq 0$, we get

$$\mathrm{div}(A(|\nabla u|)\nabla u) \leq (m-1)A(t) \cdot \left[u''(r) + \frac{N-1}{m-1} \cdot \frac{u'(r)}{1+r} \right]. \tag{2.50}$$

Since $\gamma = \frac{N-m}{m-1}$, we have $\frac{N-1}{m-1} = \gamma + 1$ and (2.49) ensures that

$$u''(r) + \frac{N-1}{m-1} \cdot \frac{u'(r)}{1+r} = -\frac{\varepsilon k(\gamma \log(1+r) + \gamma + 2)}{(1+r)^{\gamma+2} \cdot (1+\log(1+r))^3}$$

$$\leq -\frac{C_2 \varepsilon}{(1+r)^{\gamma+2} \cdot (1+\log(1+r))^2}$$

(2.51)

for some fixed constant $C_2 > 0$. In view of (2.50) and (2.51), we have

$$\mathscr{L}u = -\mathrm{div}(A(|\nabla u|)\nabla u)$$

$$\geq \frac{C_2 \varepsilon (m-1) A(t)}{(1+r)^{\gamma+2} \cdot (1+\log(1+r))^2}$$

$$\geq \frac{C_3 \varepsilon \cdot |u'(r)|^{m-1}}{(1+r)^{\gamma+2} \cdot (1+\log(1+r))^2 \cdot |u'(r)|}$$

for some fixed constant $C_3 > 0$. Using the two estimates in (2.46), we conclude that

$$\mathscr{L}u \geq \frac{C_4 \varepsilon^{m-1}(1+r)^{-\gamma(m-1)-m}}{(1+\log(1+r))^{2m}} = \frac{C_4 \varepsilon^{m-1}(1+r)^{-N}}{(1+\log(1+r))^{2m}}$$

(2.52)

for all sufficiently small $\varepsilon > 0$ because $\gamma = \frac{N-m}{m-1}$ by above.

In order to estimate the convolution term, we note that $\gamma p = \frac{p(N-m)}{m-1} > N - \alpha$ by our assumption and that $u(r) \leq \varepsilon(1+r)^{-\gamma}$ by (2.44). According to Lemma 2.12(iii), this implies

$$(|x|^{-\alpha} * u^p)u^q \leq \begin{cases} C_5 \varepsilon^{p+q}(1+r)^{N-\alpha-\gamma(p+q)} & \text{if } \gamma p < N, \\ C_5 \varepsilon^{p+q}(1+r)^{-\alpha-\gamma q}(1+\log(1+r)) & \text{if } \gamma p = N, \\ C_5 \varepsilon^{p+q}(1+r)^{-\alpha-\gamma q} & \text{if } \gamma p > N. \end{cases}$$

(2.53)

Moreover, the three conditions in part (iii) can also be expressed in the form

$$\gamma p > N - \alpha, \qquad \gamma q > N - \alpha \quad \text{and} \quad \gamma(p+q) > 2N - \alpha.$$

(2.54)

Case 1: $\gamma p < N$. Then, estimates (2.52) and (2.53) yield

$$\mathscr{L}u \geq \frac{C_4 \varepsilon^{m-1}(1+r)^{-N}}{(1+\log(1+r))^{2m}}$$

$$\geq C_5 \varepsilon^{p+q}(1+r)^{N-\alpha-\gamma(p+q)} \geq (|x|^{-\alpha} * u^p)u^q$$

for all sufficiently small $\varepsilon > 0$ because $\gamma(p+q) > 2N - \alpha$ and $p + q > m - 1$.

Case 2: $\gamma p \geq N$. In this situation, estimates (2.52) and (2.53) imply

$$\mathscr{L}u \geq \frac{C_4 \varepsilon^{m-1}(1+r)^{-N}}{(1+\log(1+r))^{2m}}$$

$$\geq C_5 \varepsilon^{p+q}(1+r)^{-\alpha-\gamma q} \cdot (1+\log(1+r)) \geq (|x|^{-\alpha} * u^p)u^q$$

for all sufficiently small $\varepsilon > 0$ because $\gamma q > N - \alpha$ and $p + q > m - 1$.
In either case, the function u given in (2.44) is a bounded radial solution of (2.1).

$\square$

From Theorem 2.15, we have a complete picture for all exponents $p > 0$, $q \in \mathbb{R}$ in case $\mathscr{A}$ is (H_m) and satisfies (2.36). These assumptions are fulfilled for the quasilinear inequalities involving the m-Laplace operator

$$-\Delta_m u \equiv -\mathrm{div}\left(|\nabla u|^{m-2}\nabla u\right) \geq (|x|^{-\alpha} * u^p)u^q \quad \text{in } \Omega \subset \mathbb{R}^N, \tag{2.55}$$

and the m-mean curvature operator

$$-\mathrm{div}\left(\frac{|\nabla u|^{m-2}}{\sqrt{1+|\nabla u|^m}}\nabla u\right) \geq (|x|^{-\alpha} * u^p)u^q \quad \text{in } \Omega \subset \mathbb{R}^N. \tag{2.56}$$

Corollary 2.16 *Assume $\mathscr{A}$ is (H_m) and satisfies (2.36) for some $m > 1$ and let $p > 0$, $q \in \mathbb{R}$ and $\alpha \in (0, N)$.*

(i) *If $m \geq N$, then (2.1) (also (2.55) and (2.56)) has no positive solutions.*
(ii) *Assume $N > m > 1$. Then, the following statements are equivalent:*

   (ii1) *Inequality (2.1) (also (2.55) and (2.56)) has a positive solution in $\mathbb{R}^N$.*
   (ii2) *Inequality (2.1) (also (2.55) and (2.56)) has a positive solution in some exterior domain.*
   (ii3) *The following conditions hold:*

   - $p + q > \dfrac{(2N-\alpha)(m-1)}{N-m}$;

   - $\begin{cases} \min\{p, q\} > \dfrac{(N-\alpha)(m-1)}{N-m} & \text{if } 0 < \alpha < m, \\[2mm] p > m-1, \, q \geq m-1 & \text{if } \alpha = m, \\[2mm] p > \dfrac{(N-\alpha)(m-1)}{N-m}, \, q > m-1-\dfrac{\alpha-m}{N}p & \text{if } m < \alpha < N. \end{cases}$

The diagrams below illustrate the existence regions in the pq-plane separately for the cases $0 < \alpha < m$ and $m \leq \alpha < N$, respectively (Figs. 2.1 and 2.2).

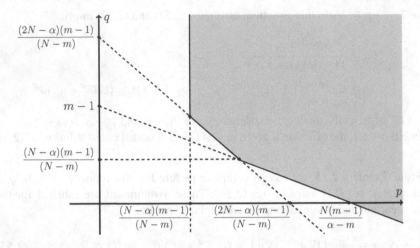

Fig. 2.1 The existence region (shaded) for positive solutions of (2.1) in the case $m < \alpha < N$

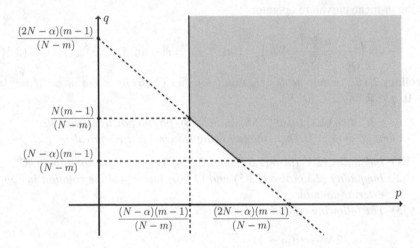

Fig. 2.2 The existence region (shaded) for positive solutions of (2.1) in the case $0 < \alpha \leq m$

2.5   Existence of Solutions in Bounded Domains

In this section, we assume that $\mathscr{A}$ satisfies (2.34) for some $A \in C[0, \infty) \cap C^1(0, \infty)$ and study inequality (2.1) in open bounded sets. Our main goal is to show that solutions exist under very mild conditions on p, q. The first results concern the case where $N \geq 2$ and

$$A(t) \geq Ct^{m-2}, \tag{2.57}$$

for $m > 1$ and some constant $C > 0$.

Theorem 2.17 *Assume $\mathscr{A}$ satisfies (2.34), $N \geq 2$, $p > 0$, $q \in \mathbb{R}$ and one of the following conditions hold:*

(i) $p + q > m - 1$ *and (2.57) holds for small enough $t > 0$.*
(ii) $p + q < m - 1$ *and (2.57) holds for large enough $t > 0$.*

Then (2.1) has a bounded radial solution in any bounded open set $\Omega \subset \mathbb{R}^N$.

Proof We assume that Ω is contained in some ball $B_R \subset \mathbb{R}^N$ for some $R > 0$.

(i) We look for solutions in the form $u(r) = \delta(1 + R - r)$ for some small enough $\delta > 0$. According to (2.37), one has

$$\text{div}(A(|\nabla u|)\nabla u) = \frac{N-1}{r} \cdot A(|u'(r)|) \cdot u'(r) = -\frac{N-1}{r} \cdot \delta A(\delta).$$

For $\delta > 0$ sufficiently small, one may employ (2.35) to find

$$\mathscr{L}u \geq \frac{N-1}{r} \cdot C_0 \delta^{m-1} \geq C_1 \delta^{m-1} \tag{2.58}$$

for all $r < R$. On the other hand, $u(r) \leq \delta(1 + R)$ by definition, so we also have

$$|x|^{-\alpha} * u^p \leq = \int_{|y|<R} \frac{u(y)^p \, dy}{|x - y|^\alpha} \leq C_2 \delta^p \int_{|y|<R} \frac{dy}{|x - y|^\alpha}.$$

Letting I_1 be the part of the integral with $|x - y| \leq |y|$ and $|y| < R$, we have

$$I_1 \leq C_2 \delta^p \int_{|x-y|<R} \frac{dy}{|x - y|^\alpha} = C_3 \delta^p.$$

Letting I_2 be the remaining part with $|x - y| \geq |y|$ and $|y| < R$, we similarly have

$$I_2 \leq C_2 \delta^p \int_{|y|<R} \frac{dy}{|y|^\alpha} = C_3 \delta^p.$$

We now combine the last two estimates to derive

$$(|x|^{-\alpha} * u^p) \cdot u^q \leq 2C_3 \delta^p \cdot u^q = 2C_3 \delta^{p+q}(1 + R - r)^q \leq C_4 \delta^{p+q}$$

for all $r < R$. Since $p + q > m - 1$ by assumption, it now follows by (2.58) that

$$\mathscr{L}u \geq C_1 \delta^{m-1} \geq C_4 \delta^{p+q} \geq (|x|^{-\alpha} * u^p) \cdot u^q$$

for all $r < R$ and all sufficiently small $\delta > 0$.

(ii) We proceed as above by taking $u(r) = L(1 + R - r)$ for some large enough $L > 0$. As above, we have

$$-\text{div}(A(|\nabla u|)\nabla u) = \frac{N-1}{r} \cdot LA(L) \geq C_1 L^{m-1}$$

for all $r < R$. On the other hand, the convolution term satisfies the estimate

$$(|x|^{-\alpha} * u^p) \cdot u^q \leq C_4 L^{p+q}$$

as before. Since $p + q < m - 1$, this yields a solution for all large enough $L > 0$.

$\square$

For $N = 1$ we have a similar statement; this time we require that instead of (2.57) the function A satisfies

$$tA'(t) + A(t) \geq Ct^{m-2}, \tag{2.59}$$

for $m > 1$ and some constant $C > 0$.

Theorem 2.18 *Assume $\mathscr{A}$ satisfies (2.34), $N = 1$, $p > 0$, $q \in \mathbb{R}$ and one of the following conditions hold:*

(i) $p + q > m - 1$ *and (2.59) holds for small enough $t > 0$.*
(ii) $p + q < m - 1$ *and (2.59) holds for large enough $t > 0$.*

Then (2.1) has a bounded radial solution in any bounded interval $\Omega \subset \mathbb{R}$.

Proof We assume that Ω is contained in the interval $(0, R)$.

(i) We look for solutions in the form $u(r) = \delta \log(1 + R + r)$ for some small enough $\delta > 0$. Since $N = 1$, we have

$$\text{div}(A(|\nabla u|)\nabla u) = u''(r) \cdot [tA'(t) + A(t)] = -\frac{\delta}{(1 + R + r)^2} \cdot [tA'(t) + A(t)],$$

where $t = |u'(r)| = \delta/(1 + R + r)$. Note that t converges to zero uniformly in Ω as $\delta \to 0$. Thus, for δ sufficiently small, we may thus conclude that

$$\mathscr{L}u = -\text{div}(A(|\nabla u|)\nabla u) \geq \frac{\delta \cdot C_0 \delta^{m-2}}{(1 + R + r)^2} \geq C_1 \delta^{m-1}$$

for all $r < R$. On the other hand, $u(r) \leq \delta \log(1 + 2R)$ by definition, so we also have

$$|x|^{-\alpha} * u^p \leq C_2 \int_{|y| < R} \frac{u(y)^p \, dy}{|x - y|^\alpha} \leq 2C_3 \delta^p.$$

Since $\log(1 + R) \leq u(r)/\delta \leq \log(1 + 2R)$, the above estimate gives

$$(|x|^{-\alpha} * u^p) \cdot u^q \leq 2C_3 \delta^p \cdot u^q \leq C_4 \delta^{p+q}$$

and the result follows as before because $p + q > m - 1$ by assumption.

(ii) We proceed as above by taking $u(r) = L \log(1 + R + r)$ for some large enough $L > 0$. Since the argument is very similar, we omit the details.

$\square$

2.6 Conclusions and Further Remarks

We investigated in this chapter some quasilinear elliptic inequalities featuring nonlocal (convolution) terms. These inequalities were posed in exterior domains or bounded domains, and we obtained (see Sect. 2.4) optimal conditions in terms of N, m, α, p, q for which positive solutions exist. Such conditions are not expressed in terms of the critical Sobolev exponent as is the case for the local equations (1.8) and (1.9). Solutions with isolated singularities for semilinear equations and inequalities with convolution terms are discussed in [CZ16, CZ18, GT15, GT16a, GT16b].

For the corresponding systems, the inequality (2.1) are studied in [GMM21] where three types of nonlocal combinations are identified:

$$\begin{cases} -\mathrm{div}[\mathscr{A}(x, u, \nabla u)] \geq (|x|^{-\alpha} * v^p) u^q \\ -\mathrm{div}[\mathscr{B}(x, v, \nabla v)] \geq (|x|^{-\beta} * u^r) v^s \end{cases} \quad \text{in } \Omega,$$

$$\begin{cases} -\mathrm{div}[\mathscr{A}(x, u, \nabla u)] \geq (|x|^{-\alpha} * v^p) v^q \\ -\mathrm{div}[\mathscr{B}(x, v, \nabla v)] \geq (|x|^{-\beta} * u^r) u^s \end{cases} \quad \text{in } \Omega,$$

$$\begin{cases} -\mathrm{div}[\mathscr{A}(x, u, \nabla u)] \geq (|x|^{-\alpha} * u^p) v^q \\ -\mathrm{div}[\mathscr{B}(x, v, \nabla v)] \geq (|x|^{-\beta} * v^r) u^s \end{cases} \quad \text{in } \Omega,$$

where $\Omega \subset \mathbb{R}^N$, $\alpha, \beta \in (0, N)$, $p, r > 0$, $q, s \in \mathbb{R}$ and $\mathscr{A}, \mathscr{B}$ are W-m_1-C and W-m_2-C respectively, for some $m_1, m_2 > 1$.

Turning back to (2.55) which is the prototype of (2.1), we mention here that a more general version of (2.55), namely,

$$-\Delta_m u - \frac{\mu}{|x|^\gamma} u^{m-1} \geq (|x|^{-\alpha} * u^p) u^q \quad \text{in } \mathbb{R}^N \setminus \overline{D}_1, \tag{2.60}$$

was studied in [GKS21] and [DGK22]. Here $p > 0$, $q, \mu \in \mathbb{R}$, $N \geq 1$, $m > 1$, $\alpha \in (0, N)$, $\gamma \leq m$. Unlike (2.55), the inequality (2.60) admits positive solutions also when $m \geq N$ provided p, q, α, γ and μ satisfy some appropriate conditions.

Chapter 3
Singular and Bounded Solutions for Quasilinear Inequalities

3.1 Introduction

In this chapter, we study the quasilinear elliptic inequality

$$\operatorname{div}\big(|x|^{-\beta}|\nabla u|^{m-2}\nabla u\big) \geq (|x|^{-\alpha} * u^p)u^q \quad \text{in } B_1 \setminus \{0\} \subset \mathbb{R}^N, \tag{3.1}$$

and the double inequality

$$a(|x|^{-\alpha} * u^p)u^q \geq \operatorname{div}\big(|x|^{-\beta}|\nabla u|^{m-2}\nabla u\big) \geq b(|x|^{-\alpha} * u^p)u^q \quad \text{in } B_1 \setminus \{0\}, \tag{3.2}$$

where $\alpha \in (0, N)$, $\beta \in \mathbb{R}$, $m > 1$, $N \geq 1$, $p > 0$, $q > m - 1$ and $a \geq b > 0$. The quantity $|x|^{-\alpha} * u^p$ represents the convolution operation as in (2.2). By a positive solution of (3.1), we understand a function $u \in W_{loc}^{1,m}(B_1 \setminus \{0\}) \cap C(\overline{B}_1 \setminus \{0\})$ which satisfies:

- $u > 0$, $u \in L^p(B_1)$, $\operatorname{div}(|x|^{-\beta}|\nabla u|^{m-2}\nabla u)$, $(|x|^{-\alpha} * u^p)u^q \in L_{loc}^1(B_1 \setminus \{0\})$;
- for any $\phi \in C_c^\infty(\Omega)$, $\phi \geq 0$, we have

$$\int_{B_1} |x|^{-\beta}|\nabla u|^{m-2}\nabla u \cdot \nabla\phi + \int_{B_1} (|x|^{-\alpha} * u^p)u^q\phi \leq 0.$$

Solutions of (3.1) are called singular if

$$\limsup_{x \to 0} u(x) = \infty.$$

Let us point out that the condition $u \in L^p(B_1)$ is needed to ensure that $|x|^{-\alpha} * u^p$ is finite almost everywhere. In fact, these two conditions are equivalent since for $x \in B_1 \setminus \{0\}$ we have

© The Author(s), under exclusive license to Springer Nature Switzerland AG 2022
M. Ghergu, *Partial Differential Inequalities with Nonlinear Convolution Terms*,
SpringerBriefs in Mathematics, https://doi.org/10.1007/978-3-031-21856-9_3

$$\infty > (|x|^{-\alpha} * u^p)(x) = C \int_{B_1} \frac{u^p(y)}{|x-y|^{\alpha}} dy \geq C \int_{B_1} \frac{u^p(y)}{2^{\alpha}} dy, \tag{3.3}$$

so $u \in L^p(B_1)$. Conversely, if $u \in L^p(B_1)$ then, by standard properties of convolution, one has $|x|^{-\alpha} * u^p \in L^1(B_1)$.

In this chapter, we study the local boundedness in B_1 of positive solutions to (3.1) as well as the existence of positive singular solutions to (3.1) and (3.2) in terms of N, m, α, β, p and q.

3.2 An L^{∞}_{loc} Regularity Criterion

The main result of this section establishes that under some conditions on the exponents, all positive solutions of (3.1) are locally bounded.

Theorem 3.1 *Assume $N > m + \beta$ and $q \geq \frac{N(m-1)}{N-m-\beta}$. Then, any positive solution u of (3.1) satisfies $u \in L^{\infty}_{loc}(B_1)$.*

We need first the following a priori estimates.

Proposition 3.2 *Let $f \in L^1_{loc}(\Omega)$, $f \geq 0$ and $\beta \in \mathbb{R}$. Suppose that $u \geq 0$ is a weak solution of*

$$\mathrm{div}(|x|^{-\beta}|\nabla u|^{m-2}\nabla u) \geq f(x) \quad in \; B_1 \setminus \{0\},$$

in the sense that

$$\int_{\Omega} |x|^{-\beta}|\nabla u|^{m-2}\nabla u \cdot \nabla \varphi + \int_{\Omega} f(x)\varphi \leq 0 \quad for\;any\;\varphi \in C^{\infty}_c(\Omega),\;\varphi \geq 0. \tag{3.4}$$

Let $\phi \in C^{\infty}_c(\Omega)$ be a standard cutoff function such that:

- *$0 \leq \phi \leq 1$ and $supp\,\phi \subset B_{4R} \setminus B_{R/2}$ for some $0 < R < 1/4$;*
- *$\phi = 1$ in $B_{2R} \setminus B_R$;*
- *$|\nabla \phi| \leq \frac{C}{R}$ in $B_1 \setminus \{0\}$.*

Then, for any $\lambda > m$ and $\ell > m - 1$, there exists $C > 0$ independent of R such that

$$\int_{\Omega} f(x)\phi^{\lambda} \leq C R^{N-m-\beta-\frac{m-1}{\ell}N} \left(\int_{\Omega} u^{\ell}\phi^{\lambda} \right)^{\frac{m-1}{\ell}}. \tag{3.5}$$

Proof Let $\theta > 0$ be small such that $\ell > (m-1)(1+\theta)$. Letting $\varphi = u^{-\theta}\phi^{\lambda}$ as a test function in (3.4), we obtain

$$\int_{\Omega} f(x)u^{-\theta}\phi^{\lambda} + \theta \int_{\Omega} |x|^{-\beta}u^{-\theta-1}\phi^{\lambda}|\nabla u|^m \leq \lambda \int_{\Omega} |x|^{-\beta}\phi^{\lambda-1}u^{-\theta}|\nabla u|^{m-1}\nabla u \cdot \nabla \phi. \tag{3.6}$$

By Young's inequality , it follows that

$$\lambda\phi^{\lambda-1}u^{-\theta}|\nabla u|^{m-1}\nabla u \cdot \nabla\phi \le \frac{\theta}{2}u^{-\theta-1}\phi^\lambda|\nabla u|^m + C(\theta,\lambda)u^{m-1-\theta}\phi^{\lambda-m}|\nabla\phi|^m.$$

Using the above inequality in (3.6), we find

$$\int_\Omega f(x)u^{-\theta}\phi^\lambda + \frac{\theta}{2}\int_\Omega |x|^{-\beta}u^{-\theta-1}\phi^\lambda|\nabla u|^m \le C\int_\Omega |x|^{-\beta}u^{m-1-\theta}\phi^{\lambda-m}|\nabla\phi|^m.$$

In particular, this yields

$$\int_\Omega |x|^{-\beta}u^{-\theta-1}\phi^\lambda|\nabla u|^m \le CR^{-\beta}\int_\Omega u^{m-1-\theta}\phi^{\lambda-m}|\nabla\phi|^m. \tag{3.7}$$

Let now $\varphi = \phi^\lambda$ in (3.4). By Hölder's inequality, we have

$$\int_\Omega f(x)\phi^\lambda \le \lambda\int_\Omega |x|^{-\beta}\phi^{\lambda-1}|\nabla u|^{m-2}\nabla u \cdot \nabla\phi$$

$$\le \lambda\Big(\int_\Omega |x|^{-\beta}u^{-\theta-1}\phi^\lambda|\nabla u|^m\Big)^{1/m'}\Big(\int_\Omega |x|^{-\beta}u^{(1+\theta)(m-1)}\phi^{\lambda-m}|\nabla\phi|^m\Big)^{1/m}.$$

In particular, we have

$$\int_\Omega f(x)\phi^\lambda \le CR^{-\beta/m}\Big(\int_\Omega |x|^{-\beta}u^{-\theta-1}\phi^\lambda|\nabla u|^m\Big)^{1/m'}\Big(\int_\Omega u^{(1+\theta)(m-1)}\phi^{\lambda-m}|\nabla\phi|^m\Big)^{1/m}, \tag{3.8}$$

which combined with (3.7) yields

$$\int_\Omega f(x)\phi^\lambda \le CR^{-\beta}\Big(\int_\Omega u^{m-1-\theta}\phi^{\lambda-m}|\nabla\phi|^m\Big)^{1/m'}\Big(\int_\Omega u^{(1+\theta)(m-1)}\phi^{\lambda-m}|\nabla\phi|^m\Big)^{1/m}. \tag{3.9}$$

We next use the Hölder's inequality with exponents

$$\gamma = \frac{\ell}{m-1} > 1 \quad \text{and} \quad \tau = \frac{\ell}{(m-1)(\theta+1)} > 1$$

to obtain

$$\int_\Omega u^{m-1-\theta}\phi^{\lambda-m}|\nabla\phi|^m \le \Big(\int_\Omega u^\ell\phi^\lambda\Big)^{1/\gamma}\Big(\int_\Omega \phi^{\lambda-m\gamma'}|\nabla\phi|^{m\gamma'}\Big)^{1/\gamma'} \tag{3.10}$$

$$\le CR^{\frac{N}{\gamma'}-m}\Big(\int_\Omega u^\ell\phi^\lambda\Big)^{1/\gamma},$$

and

$$\int_\Omega u^{(1+\theta)(m-1)} \phi^{\lambda-m} |\nabla\phi|^m \le \left(\int_\Omega u^\ell \phi^\lambda \right)^{1/\tau} \left(\int_\Omega \phi^{\lambda-m\tau'} |\nabla\phi|^{m\tau'} \right)^{1/\tau'}$$

(3.11)

$$\le C R^{\frac{N}{\tau'}-m} \left(\int_\Omega u^\ell \phi^\lambda \right)^{1/\tau}.$$

Now, we use (3.10) and (3.11) into (3.9) to derive the conclusion. □

Before we prove Theorem 3.1, let us first establish the following asymptotic behaviour of positive solutions related to the local version of (3.1):

$$\mathrm{div}(|x|^{-\beta}|\nabla u|^{m-2}\nabla u) \ge |x|^{-\theta} u^q \quad \text{in } B_1 \setminus \{0\}.$$

(3.12)

Proposition 3.3 *Let $\theta \ge 0$ and $q > \max\{m-1, \theta\}$.*
If $u \in W_{loc}^{1,m}(B_1 \setminus \{0\}) \cap C(\overline{B}_1 \setminus \{0\})$ is a positive solution of (3.12), then

$$u(x) \le C|x|^{-\frac{m+\beta-\theta}{q-m+1}} \quad \text{for all } x \in B_1 \setminus \{0\},$$

(3.13)

for some constant $C > 0$.

Proof We apply Proposition 3.2 with $\ell = q > m-1$ and $\gamma = 0$. According to the estimate (3.5), we have

$$C R^{N-m-\beta-\frac{m-1}{q}N} \left(\int_{B_1} u^q \phi^\lambda \right)^{\frac{m-1}{q}} \ge \int_{B_1} |x|^{-\theta} u^q \phi^\lambda.$$

Since $\mathrm{supp}\,\phi \subset B_{4R} \setminus B_{R/2}$, from the above estimate and the weak Harnack inequality (B.6) with $a = 7/4$, $b = 5/4$ and $c = 1/8$, it follows that

$$C R^{N-m-\beta-\frac{m-1}{q}N} \ge R^{-\theta} \left(\int_{B_{2R} \setminus B_R} u^q \right)^{1-\frac{m-1}{q}}$$

$$\ge R^{-\theta} \left(R^N \sup_{\frac{5R}{4} < |x| < \frac{7R}{4}} u^q \right)^{1-\frac{m-1}{\ell}}$$

$$\ge R^{N-\theta-\frac{m-1}{q}N} \left(\sup_{\frac{5R}{4} < |x| < \frac{7R}{4}} u^{q-m+1} \right).$$

From here, we easily deduce (3.13). □

Proof of Theorem 3.1 Let $\nu = \frac{N(m-1)}{N-m-\beta}$ and let u be a positive solution of (3.1). We note that since $u \in L^p(B_1)$, u satisfies

$$\mathrm{div}(|x|^{-\beta}|\nabla u|^{m-2}\nabla u) \ge c\, u^q \quad \text{in } B_1 \setminus \{0\},$$

(3.14)

where $c = 2^{\beta-N} \int_{B_1} u^p > 0$, by (3.3). Using Proposition 3.3 (with $\theta = 0$ and since $q \geq \nu > m - 1$), we deduce

$$u(x) \leq C|x|^{-\frac{m+\beta}{q-m+1}} \quad \text{in } B_1 \setminus \{0\}.$$

In particular, again by $q \geq \nu$, it follows that

$$u(x) \leq C|x|^{-\frac{m+\beta}{\nu-m+1}} \quad \text{in } B_1 \setminus \{0\}. \tag{3.15}$$

Also, from (3.14) we deduce

$$\text{div}(|x|^{-\beta}|\nabla u|^{m-2}\nabla u) \geq cu^\nu - C \quad \text{in } B_1 \setminus \{0\}, \tag{3.16}$$

for some $C > 0$. $\qquad\qquad\square$

In order to proceed to the proof of Theorem 3.1, we need the following result.

Lemma 3.4 *Assume u satisfies (3.16). Then, for any $\phi \in C_c^1(B_1 \setminus \{0\})$, $\phi \geq 0$ and any number $M \geq (C/c)^\nu$, we have*

$$\left\| |x|^{-\beta/m}\phi|\nabla(u-M)^+| \right\|_{L^m(B_1)} \leq m \left\| |x|^{-\beta/m}(u-M)^+|\nabla\phi| \right\|_{L^m(B_1)}. \tag{3.17}$$

Proof of Lemma 3.4 Let $\{\eta_k\} \subset C^1(\mathbb{R})$ be such that $\eta_k \geq 0$,

$$\eta'_k = 0 \text{ on } (-\infty, 0), \quad \eta'_k > 0 \text{ on } (0, \infty),$$

$$\eta'_k(t) \to \text{sign}^+(t), \quad \eta_k(t) \to t^+ \quad \text{as } k \to \infty,$$

where $\text{sign}^+(t) = 1$ if $t > 0$ and $\text{sign}^+(t) = 0$ if $t < 0$. Take $\eta_k(u - M)\phi^m$ as a test function in (3.16). We find

$$\int_{B_1} (cu^\nu - C)\eta_k(u-M)\phi^m dx + \int_{B_1} |x|^{-\beta}|\nabla u|^{m-2}\nabla u \cdot \nabla\big(\eta_k(u-M)\phi^m\big)dx \leq 0.$$

Since $(cu^\nu - C)\eta_k(u-M)\phi^m \geq 0$, by the choice of M, it follows that

$$\int_{B_1} |x|^{-\beta}\Big[\phi^m \eta'_k(u-M)|\nabla u|^{m-2}\nabla u \cdot \nabla(u-M)^+ + m\phi^{m-1}\eta_k(u-M)|\nabla u|^{m-2}\nabla u \cdot \nabla\phi\Big]dx \leq 0.$$

Letting $k \to \infty$, by Fatou's lemma, we find

$$\int_{B_1} |x|^{-\beta}\phi^m|\nabla(u-M)^+|^m dx + m\int_{B_1} |x|^{-\beta}\phi^{m-1}(u-M)^+|\nabla u|^{m-2}\nabla u \cdot \nabla\phi dx \leq 0,$$

so

$$\int_{B_1} |x|^{-\beta} \phi^m |\nabla(u-M)^+|^m dx \leq m \int_{B_1} |x|^{-\beta} \phi^{m-1} (u-M)^+ |\nabla u|^{m-1} |\nabla \phi| dx. \qquad (3.18)$$

By Hölder's inequality and since $(u-M)^+ |\nabla u| = (u-M)^+ |\nabla(u-M)^+|$, we estimate the right-hand side of (3.18) as

$$\int_{B_1} |x|^{-\beta} \phi^{m-1} (u-M)^+ |\nabla u|^{m-1} |\nabla \phi| dx \leq \left(\int_{B_1} |x|^{-\beta} \phi^m |\nabla(u-M)^+|^m dx \right)^{1/m'} \times$$

$$\times \left(\int_{B_1} |x|^{-\beta} |\nabla \phi|^m |(u-M)^+|^m dx \right)^{1/m},$$

$$(3.19)$$

where m' is the Hölder conjugate of m. Using (3.19) into (3.18), we deduce (3.17). $\qquad \square$

We are now ready to proceed to the proof of Theorem 3.1 whose arguments will be divided into two steps.

Step 1: $u \in L^\nu_{loc}(B_1)$. Let $\eta \in C^1(\mathbb{R})$ be such that $\eta \geq 0$, η is bounded, $\eta = 0$ on $(-\infty, 0)$ and $\eta' > 0$ on $(0, \infty)$. Let also $\{\zeta_k\} \in C^1_c(\mathbb{R}^N)$ be such that

$$\zeta_k(x) = \begin{cases} 0 & \text{if } |x| < \dfrac{1}{2k} \text{ or } |x| > \dfrac{2}{3}, \\ 1 & \text{if } \dfrac{1}{k} < |x| < \dfrac{1}{2}, \end{cases} \quad \text{and} \quad |\nabla \zeta_k| \leq Ck.$$

Define $A_k = B_{1/k} \setminus B_{1/(2k)}$ and

$$M \geq \max \left\{ (C/c)^q, \max_{1/2 \leq |x| \leq 2/3} u(x) \right\}.$$

We next test (3.16) with $\zeta_k \eta(u-M)$. We find

$$\int_{B_1} (cu^\nu - C) \zeta_k \eta(u-M) dx + \int_{B_1} |x|^{-\beta} |\nabla u|^{m-2} \nabla u \cdot \nabla \big(\zeta_k \eta(u-M) \big) dx \leq 0.$$

Since $\eta' \geq 0$ and $|\nabla u|^{m-2} \nabla u \nabla(u-M) = |\nabla u|^m \geq 0$, it follows that

$$\int_{B_1} (cu^\nu - C) \zeta_k \eta(u-M) dx \leq \Gamma_k := \int_{B_1} |x|^{-\beta} \eta(u-M) |\nabla u|^{m-1} |\nabla \zeta_k| dx.$$

$$(3.20)$$

Observe that $\eta(u-M) \nabla \zeta_k = 0$ outside of A_k, since $M \geq \max_{1/2 \leq |x| \leq 2/3} u(x)$. Using the fact that η is bounded together with Hölder's inequality, we find

$$\Gamma_k \le \|\eta\|_\infty \int_{A_k} |x|^{-\beta} |\nabla(u-M)^+|^{m-1} |\nabla\zeta_k| dx$$

$$\le C \left\| |x|^{-\beta/m} |\nabla(u-M)^+| \right\|_{L^m(A_k)}^{m-1} \left\| |x|^{-\beta/m} |\nabla\zeta_k| \right\|_{L^m(A_k)}. \tag{3.21}$$

By the definition of ζ_k and the fact that $|\nabla\zeta_k| \le ck$, we have

$$\left\| |x|^{-\beta/m} |\nabla\zeta_k| \right\|_{L^m(A_k)} \le Ck^{1-\frac{N-\beta}{m}}.$$

Using this fact in (3.21) together with $\zeta_{2k} = 1$ in A_k and $\zeta_k \ge 0$ in A_{2k}, we further estimate

$$\Gamma_k \le Ck^{1-\frac{N-\beta}{m}} \left\| |x|^{-\beta/m} |\nabla(u-M)^+| \right\|_{L^m(A_k)}^{m-1}$$

$$\le Ck^{1-\frac{N-\beta}{m}} \left\| |x|^{-\beta/m} \zeta_{2k} |\nabla(u-M)^+| \right\|_{L^m(A_{2k}\cup A_k)}^{m-1} \tag{3.22}$$

$$\le Ck^{1-\frac{N-\beta}{m}} \left\| |x|^{-\beta/m} (u-M)^+ |\nabla\zeta_{2k}| \right\|_{L^m(A_{2k})}^{m-1},$$

where in the last inequality, we have used (3.17) with $\phi = \zeta_{2k}$ and the fact that $\nabla\zeta_{2k} = 0$ in A_k. From (3.15) we have

$$\int_{A_{2k}} |x|^{-\beta} |(u-M)^+|^m |\nabla\zeta_{2k}|^m dx \le Ck^{\beta+m} \int_{A_{2k}} |(u-M)^+|^m \le Ck^{\beta+m-N+\frac{m(m+\beta)}{\nu-m+1}}.$$

Hence, from (3.22) we deduce

$$\Gamma_k \le Ck^{1-\frac{N-\beta}{m}+\left(\beta+m-N+\frac{m(m+\beta)}{\nu-m+1}\right)\frac{m-1}{m}} = C.$$

We now replace η in (3.20) by a sequence $\{\eta_n\}$ such that $\eta_n(t) \to \mathrm{sign}^+(t)$ as $n \to \infty$. Letting $n \to \infty$ and then $k \to \infty$ in (3.20), since supp $\zeta_k = \overline{B}_{2/3}$ and $\zeta_k \to 1$ in $B_{1/2}$, we find

$$\int_{B_1} (cu^\nu - C)\mathrm{sign}^+(u-M)dx \le C,$$

so $u \in L_{loc}^\nu(B_1)$.

Step 2: $u \in L_{loc}^\infty(B_1)$. We return to the estimate (3.22) and split our analysis into two cases.

- Case 2.1: $\nu \ge m$. By Hölder's inequality, we find

$$\left\||x|^{-\beta/m}(u-M)^+|\nabla\zeta_{2k}|\right\|_{L^m(A_{2k})} \le Ck^{\frac{\beta}{m}+1}\left\|(u-M)^+\right\|_{L^m(A_{2k})}$$

$$\le Ck^{\frac{\beta}{m}+1}\left\|(u-M)^+\right\|_{L^\nu(A_{2k})}|A_{2k}|^{\frac{1}{m}-\frac{1}{\nu}}$$

$$= Ck^{\frac{\beta}{m}+1-N\left(\frac{1}{m}-\frac{1}{\nu}\right)}o(1) \text{ as } k \to \infty.$$

Using this estimate in (3.22), we deduce $\Gamma_k \le k^{\frac{N(m-1)}{\nu}-N+m+\beta}o(1) = o(1)$ as $k \to \infty$, thanks to the value of ν.

- Case 2.2: $\nu < m$. From (3.15) we have

$$\left\||x|^{-\beta/m}(u-M)^+|\nabla\zeta_{2k}|\right\|_{L^m(A_{2k})} \le Ck^{\frac{\beta}{m}+1}\left\|(u-M)^+\right\|_{L^m(A_{2k})}$$

$$\le Ck^{\frac{\beta}{m}+1}\sup_{A_{2k}}|(u-M)^+|^{1-\frac{\nu}{m}}\left\|(u-M)^+\right\|_{L^\nu(A_{2k})}^{\frac{\nu}{m}}$$

$$= Ck^{\frac{\beta}{m}+1}\sup_{A_{2k}}|(u-M)^+|^{1-\frac{\nu}{m}}o(1)$$

$$\le Ck^{\frac{\beta}{m}+1+\frac{m+\beta}{\nu-m+1}\left(1-\frac{\nu}{m}\right)}o(1) \text{ as } k \to \infty,$$

and from (3.22) we again derive $\Gamma_k \le Ck^{[N(m-1)-\nu(N-m-\beta)]/m}o(1) = o(1)$ as $k \to \infty$, thanks to the value of ν.

We now return to (3.20) and let $k \to \infty$ to deduce

$$\int_{B_{1/2}} (cu^\nu - C)\eta(u-M)dx = 0.$$

Since $\eta \ge 0$, it follows that $u \le M$ in $B_{1/2}$, so $u \in L^\infty_{loc}(B_1)$ which completes our proof.   □

3.3 Keller-Osserman Estimates

Let also

$$\Phi_{m,\beta}(x) = \begin{cases} |x|^{-\frac{N-m-\beta}{m-1}} & \text{if } N \ne m+\beta, \\ \log\dfrac{5}{|x|} & \text{if } N = m+\beta, \end{cases} \tag{3.23}$$

be the fundamental solution of the weighted m-Laplace operator $-\text{div}\big(|x|^{-\beta}|\nabla u|^{m-2}\nabla u\big)$ for $m > 1$. Note that $\Phi_{m,\beta}$ satisfies the distributional equality

$$-\text{div}\big(|x|^{-\beta}|\nabla\Phi_{m,\beta}|^{m-2}\nabla\Phi_{m,\beta}\big) = c\delta_0 \quad \text{in } \mathscr{D}'(\mathbb{R}^N),$$

for some positive constant c, where δ_0 denotes the Dirac delta measure.

Proposition 3.5 *Assume* $u \in W^{1,m}_{loc}(B_1 \setminus \{0\}) \cap C(B_1 \setminus \{0\})$ *satisfies* $u \geq 0$ *and*

$$\mathrm{div}(|x|^{-\beta}|\nabla u|^{m-2}\nabla u) \geq 0 \quad in \ B_1 \setminus \{0\}.$$

Then, either u is bounded near the origin, or there exist $C > 0$ *and* $r_0 \in (0, 1/2)$ *such that*

$$\sup_{|x|=r} \frac{u(x)}{\Phi_{m,\beta}(x)} \geq C \quad for \ all \ r \in (0, r_0), \tag{3.24}$$

where $\Phi_{m,\beta}$ *is defined in (3.23).*

Proof Assume that (3.24) does not hold. Hence,

$$\liminf_{r \to 0} \left(\sup_{|x|=r} \frac{u(x)}{\Phi_{m,\beta}(x)} \right) = 0.$$

Then, for any $k \geq 1$, there exists $r_k \in (0, 1/2)$, with $r_k \to 0$ as $k \to \infty$, such that

$$\sup_{|x|=r_k} \frac{u(x)}{\Phi_{m,\beta}(x)} \leq \frac{1}{k} \quad for \ all \ k \geq 1.$$

A comparison principle in the annular region $B_{1/2} \setminus B_{r_k}$ shows that for all $k \geq 1$, we have

$$u(x) \leq \frac{1}{k}\Phi_{m,\beta}(x) + \max_{|x|=1/2} u(x) \quad in \ B_{1/2} \setminus B_{r_k},$$

Letting $k \to \infty$ in the above estimate, we deduce that u is bounded in the ball $B_{1/2}$.
□

The result below provides a first important estimate for solutions to (3.1).

Proposition 3.6 (Keller-Osserman-Type Estimates) *Assume* $p + q > 2(m - 1)$ *and let* $u \in W^{1,m}_{loc}(B_1 \setminus \{0\}) \cap C(B_1 \setminus \{0\})$ *be a positive solution of (3.1). Then, there exist* $C > 0$ *such that*

$$u(x) \leq C|x|^{-\sigma} \quad in \ B_1 \setminus \{0\}, \tag{3.25}$$

where $\sigma > 0$ *is given by*

$$\sigma = \frac{N + m - \alpha + \beta}{p + q - m + 1} > 0. \tag{3.26}$$

Proof We use Proposition 3.2 with $\Omega = B_1$, $R \in (0, 1/4)$, $\gamma = 0$, $f(x) = (|x|^{-\alpha} * u^p)u^q$ and $\ell = (p + q)/2 > m - 1$. From (3.5) we find

$$CR^{N-m-\beta-\frac{m-1}{\ell}N}\left(\int_{B_1} u^\ell \phi^\lambda\right)^{\frac{m-1}{\ell}} \geq \int_{B_1} (|x|^{-\alpha} * u^p)u^q \phi^\lambda, \qquad (3.27)$$

where $\phi \in C_c^\infty(B_1 \setminus \{0\})$ and $\lambda > m$ are chosen as in Proposition 3.2. If $x, y \in B_{2R} \subset \mathrm{supp}\,\phi$, then $|x - y| \leq |x| + |y| \leq 4R$, so

$$(|x|^{-\alpha} * u^p)(x) \geq C \int_{B_{4R}} \frac{u^p(y)}{|x-y|^\alpha}dy \geq C \int_{B_{2R}} \frac{u^p(y)}{(4R)^\alpha}dy$$

$$\geq CR^{-\alpha} \int_{B_1} u^p(y)\phi^\lambda(y)dy,$$

since $0 \leq \phi \leq 1$. Using this fact in (3.27) together with Hölder's inequality, for $\ell = (p+q)/2$, we find

$$CR^\tau \left(\int_{B_1} u^\ell \phi^\lambda\right)^{\frac{m-1}{\ell}} \geq \left(\int_{B_1} u^p \phi^\lambda\right)\left(\int_{B_1} u^q \phi^\lambda\right) \geq \left(\int_{B_1} u^\ell \phi^\lambda\right)^2,$$

where

$$\tau = N - m + \alpha - \beta - \frac{m-1}{\ell}N. \qquad (3.28)$$

Now, using the fact that $\phi = 1$ in $B_{2R} \setminus B_R$ and the weak Harnack inequality (B.6) with $a = 7/4$, $b = 5/4$ and $c = 1/8$, we deduce

$$CR^\tau \geq \left(\int_{B_1} u^\ell \phi^\lambda\right)^{2-\frac{m-1}{\ell}} \geq \left(\int_{B_{2R}\setminus B_R} u^\ell\right)^{2-\frac{m-1}{\ell}}$$

$$\geq \left(R^N \sup_{\frac{5R}{4}<|x|<\frac{7R}{4}} u^\ell\right)^{2-\frac{m-1}{\ell}} \geq R^{2N-\frac{m-1}{\ell}N}\left(\sup_{\frac{5R}{4}<|x|<\frac{7R}{4}} u^{p+q-m+1}\right).$$

From here and (3.28), we derive (3.25). $\qquad\qquad\qquad\qquad\qquad\qquad\qquad\qquad \square$

3.4   Singular Solutions

This section is concerned with the study of positive solutions to (3.1) which are singular around the origin. The main result of this section is stated below.

Theorem 3.7 *Assume* $m > 1$, $N \geq 1$, $q > m - 1$, $\alpha \in (0, N)$ *and* $\beta > -m - \alpha$.

 (i) *If* $N \leq m + \beta$, *then* (3.1) *has always singular solutions.*

(ii) *If* $N > m + \beta$ *and* $p > m - 1$, *then* (3.1) *has singular solutions if and only if*

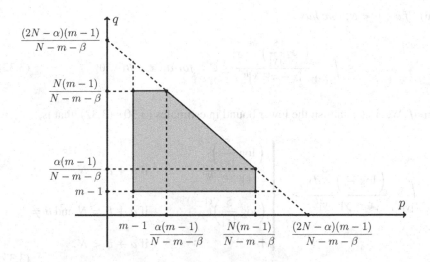

Fig. 3.1 The existence of singular solutions region to (3.1) in the pq plane

$$\max\{p,q\} < \frac{N(m-1)}{N-m-\beta}, \quad p+q < \frac{(2N-\alpha)(m-1)}{N-m-\beta} \quad and \quad \alpha < 2(m+\beta).$$
(3.29)

The picture above illustrates the region of existence in the pq-plane of singular solutions in the case $N > m + \beta$, $p > m - 1$ and $\alpha < 2(m + \beta)$ (Fig. 3.1).

Before we prove the above result, we provide some estimates for the convolution with logarithmic terms in the unit ball.

Lemma 3.8 *Let $a, b \in (0, N)$ and $\theta \geq 0$. Then, there exists $C > c > 0$ such that:*

(i) *If $a + b > N$, one has*

$$\frac{c\left(\log\frac{5}{|x|}\right)^{-\theta}}{|x|^{a+b-N}} \leq \int_{|y|<1} \frac{\left(\log\frac{5}{|y|}\right)^{-\theta} dy}{|x-y|^a|y|^b} \leq \frac{C\left(\log\frac{5}{|x|}\right)^{-\theta}}{|x|^{a+b-N}} \quad for\ all\ x \in B_1 \setminus \{0\}.$$
(3.30)

(ii) *If $a + b = N$, $\theta \neq 1$, one has*

$$c\left(\log\frac{5}{|x|}\right)^{(1-\theta)^+} \leq \int_{|y|<1} \frac{\left(\log\frac{5}{|y|}\right)^{-\theta} dy}{|x-y|^a|y|^b} \leq C\left(\log\frac{5}{|x|}\right)^{(1-\theta)^+} \quad for\ all\ x \in B_1 \setminus \{0\},$$
(3.31)

(iii) *If $a + b < N$, one has*

$$c \le \int_{|y|<1} \frac{\left(\log \frac{5}{|y|}\right)^{-\theta} dy}{|x - y|^a |y|^b} \le C \quad \text{for all } x \in B_1 \setminus \{0\}. \tag{3.32}$$

Proof We first establish the lower bound in estimates (3.30)–(3.32), that is,

$$\int_{|y|<1} \frac{\left(\log \frac{5}{|y|}\right)^{-\theta} dy}{|x - y|^a |y|^b} \ge c \begin{cases} \dfrac{\left(\log \frac{5}{|x|}\right)^{-\theta}}{|x|^{a+b-N}} & \text{if } a + b > N, \\[3ex] \left(\log \frac{5}{|x|}\right)^{(1-\theta)^+} & \text{if } a + b = N \text{ and } \theta \ne 1, \\[2ex] 1 & \text{if } a + b < N. \end{cases} \tag{3.33}$$

It is enough to establish (3.33) for all $x \in B_{1/2} \setminus \{0\}$. Then, since all functions in (3.33) are continuous and positive on $\overline{B}_1 \setminus B_{1/2}$, we may take a smaller constant $c > 0$ such that (3.33) still holds for all $x \in B_1 \setminus \{0\}$.

Observe that

$$\int_{|y|<1} \frac{\left(\log \frac{5}{|y|}\right)^{-\theta} dy}{|x - y|^a |y|^b} \ge \int_{|x|<|y|<1} \frac{\left(\log \frac{5}{|y|}\right)^{-\theta} dy}{|x - y|^a |y|^b}$$

$$\ge \int_{|x|<|y|<1} \frac{\left(\log \frac{5}{|y|}\right)^{-\theta} dy}{(2|y|)^a |y|^b}$$

$$= \sigma_N 2^{-a} \int_{|x|}^1 t^{N-a-b} \left(\log \frac{5}{t}\right)^{-\theta} \frac{dt}{t},$$

where σ_N is the surface area of the unit ball in $\mathbb{R}^N$. From here we estimate as follows:

(i1) If $a + b > N$, then

$$\int_{|x|}^1 t^{N-a-b} \left(\log \frac{5}{t}\right)^{-\theta} \frac{dt}{t} \ge \left(\log \frac{5}{|x|}\right)^{-\theta} \int_{|x|}^1 t^{N-a-b} \frac{dt}{t} \ge \frac{c \left(\log \frac{5}{|x|}\right)^{-\theta}}{|x|^{a+b-N}},$$

if $0 < |x| < 1/2$.

(i2) If $a + b = N$, then for any $0 < |x| < 1/2$, we have

$$\int_{|x|}^{1} t^{N-a-b}\left(\log\frac{5}{t}\right)^{-\theta}\frac{dt}{t} = \int_{|x|}^{1}\left(\log\frac{5}{t}\right)^{-\theta}\frac{dt}{t}$$

$$= \frac{1}{1-\theta}\left[\left(\log\frac{5}{|x|}\right)^{1-\theta} - \left(\log 5\right)^{1-\theta}\right]$$

$$\geq c\begin{cases}\left(\log\dfrac{5}{|x|}\right)^{1-\theta} & \text{if } 0 \leq \theta < 1, \\ 1 & \text{if } \theta > 1.\end{cases}$$

(i3) If $a+b < N$, and $0 < |x| < 1/2$, we have

$$\int_{|x|}^{1} t^{N-a-b}\left(\log\frac{5}{t}\right)^{-\theta}\frac{dt}{t} \geq (\log 10)^{-\theta}\int_{1/2}^{1} t^{N-a-b}\frac{dt}{t} = c(N, a, b, \theta) > 0.$$

To establish the upper bounds in estimates (3.30)–(3.32), we let $r = |x| \in (0, 1)$ and use the the change of variables $x = r\zeta$, $y = r\eta$. In particular, we have $|\zeta| = 1$. Thus

$$\int_{|y|<1}\frac{\left(\log\frac{5}{|y|}\right)^{-\theta}dy}{|x-y|^a|y|^b} \leq \int_{|y|<2}\frac{\left(\log\frac{5}{|y|}\right)^{-\theta}dy}{|x-y|^a|y|^b} = r^{N-a-b}\int_{|\eta|<2/r}\frac{\left(\log\frac{5}{r|\eta|}\right)^{-\theta}d\eta}{|\zeta-\eta|^a|\eta|^b}$$

$$\leq r^{N-a-b}\left[\left(\log\frac{5}{2r}\right)^{-\theta}\int_{0<|\eta|<2}\frac{d\eta}{|\zeta-\eta|^a|\eta|^b} + \int_{2<|\eta|<2/r}\frac{\left(\log\frac{5}{r|\eta|}\right)^{-\theta}d\eta}{|\zeta-\eta|^a|\eta|^b}\right]$$

$$\leq r^{N-a-b}\left[A\left(\log\frac{5}{2r}\right)^{-\theta} + \int_{2<|\eta|<2/r}\frac{\left(\log\frac{5}{r|\eta|}\right)^{-\theta}d\eta}{(|\eta|/2)^a|\eta|^b}\right]$$

where

$$A = \max_{|\zeta|=1}\int_{0<|\eta|<2}\frac{d\eta}{|\zeta-\eta|^a|\eta|^b} \in (0, \infty),$$

and where we have used the trivial inequality

$$|\zeta - \eta|^a \geq ||\zeta| - |\eta||^a = (|\eta| - 1)^a \geq (|\eta|/2)^a.$$

Thus

$$\int_{|y|<1}\frac{\left(\log\frac{5}{|y|}\right)^{-\theta}dy}{|x-y|^a|y|^b} \leq Cr^{N-a-b}\left[\left(\log\frac{5}{r}\right)^{-\theta} + \int_{2}^{2/r} t^{N-a-b}\left(\log\frac{5}{rt}\right)^{-\theta}\frac{dt}{t}\right].$$

Next, a straightforward calculation leads to the desired estimates in the upper bounds of (i)–(iii).

$\square$

Remark 3.9 A direct and useful calculation shows that if

$$u(x) = \kappa |x|^{-\gamma} \left(\log \frac{5}{|x|} \right)^{-\tau}, \gamma > 0,$$

then

$$\mathrm{div}\big(|x|^{-\beta} |\nabla u|^{m-2} \nabla u\big) = \kappa^{m-1} |x|^{-\gamma(m-1)-m-\beta} \left(\log \frac{5}{|x|} \right)^{-\tau(m-1)-m} \times$$

$$\times \left| -\gamma \log \frac{5}{|x|} + \tau \right|^{m-2} \left[A \left(\log \frac{5}{|x|} \right)^2 + B \log \frac{5}{|x|} + C \right],$$

where

$$A = \gamma \big[\gamma(m-1) - (N-m-\beta) \big],$$
$$B = \tau \big[-2\gamma(m-1) + (N-m-\beta) \big], \tag{3.34}$$
$$C = (m-1)\tau(\tau+1).$$

Proof of Theorem 3.7 Given two positive functions f, g defined on $\overline{B}_1 \setminus \{0\}$, we use the symbol $f \asymp g$ to denote that the quotient f/g is bounded on $\overline{B}_1 \setminus \{0\}$ between two positive constants.

(i) Let

$$0 < \gamma < \min \left\{ \frac{N-\alpha}{p}, \frac{m+\beta}{q-m+1} \right\}. \tag{3.35}$$

We show that $u(x) = \kappa |x|^{-\gamma}$ is a singular radially symmetric solution of (3.1) for suitable $\kappa \in (0, 1)$. Since from (3.35) we have $p\gamma < N - \alpha < N$, it follows that $u \in L^p(B_1)$. By Remark 3.9 (in which we take $\tau = 0$), one has

$$\mathrm{div}\big(|x|^{-\beta} |\nabla u|^{m-2} \nabla u\big) = \kappa^{m-1} \gamma^{m-2} A |x|^{-(m-1)\gamma - m - \beta} \quad \text{in } B_1 \setminus \{0\}, \tag{3.36}$$

where A is defined in $(3.34)_1$. From $N \leq m + \beta$ and $\gamma > 0$, we have $A > 0$. Further, since $p\gamma < \alpha$, by Lemma 3.8(iii) with $\theta = 0$, $a = \alpha$, $b = p\gamma$ and $a + b < N$, we estimate

$$(|x|^{-\alpha} * u^p) u^q(x) \asymp \kappa^p u^q(x) = \kappa^{p+q} |x|^{-\gamma q} \quad \text{in } B_1 \setminus \{0\}. \tag{3.37}$$

Comparing (3.36) and (3.37), we see that for $\kappa \in (0, 1)$ small enough, thanks to (3.35) and $q > m - 1$, one has that $u(x) = \kappa |x|^{-\gamma}$ is a singular positive solution of (3.1).

(ii) Let u be a positive singular solution of (3.1). Using Propositions 3.5 and 3.6, there exists $C > 0$ such that for small $R > 0$, we find

$$CR^{-\sigma} \geq \sup_{|x|=R} u \geq cR^{-\frac{N-m-\beta}{m-1}}, \tag{3.38}$$

where $\sigma > 0$ is defined in (3.26). The above estimate implies $\sigma \geq \frac{N-m-\beta}{m-1}$ which is equivalent to

$$p + q \leq \frac{(2N - \alpha)(m - 1)}{N - m - \beta}.$$

We claim that the above inequality is strict. Assuming the contrary, we have that $\sigma = \frac{N-m-\beta}{m-1}$. Let $x \in B_{1/2} \setminus \{0\}$. Combining the estimate (3.38) with the weak Harnack inequality (B.6) with $a = 1, b = 1/2, c = 1/8$ and $p > m - 1$, we find

$$(|x|^{-\alpha} * u^p)(x) \geq \int_{B_{5|x|/4} \setminus B_{|x|/4}} \frac{u^p(y)}{|x - y|^\alpha} dy$$

$$\geq \left(\frac{9|x|}{4}\right)^{-\alpha} \int_{B_{5|x|/4} \setminus B_{|x|/4}} u^p(y) dy$$

$$\geq C|x|^{-\alpha} \left(|x|^{N/p} \sup_{\partial B_{|x|}} u\right)^p \qquad \text{(by the Harnack inequality (B.6))}$$

$$\geq C|x|^{N-\alpha-\sigma p} \qquad \text{(by estimate (3.38))}.$$

Hence, u satisfies

$$\text{div}\left(|x|^{-\beta}|\nabla u|^{m-2}\nabla u\right) \geq C|x|^{N-\alpha-\sigma p}u^q \qquad \text{in } B_{1/2} \setminus \{0\}.$$

For any $k \geq 3$, let $v_k \in C^1(B_{1/2} \setminus B_{1/k})$ be a radial function such that

$$\begin{cases} \text{div}\left(|x|^{-\beta}|\nabla v_k|^{m-2}\nabla v_k\right) = C|x|^{N-\alpha-\sigma p}v_k^q & \text{in } B_{1/2} \setminus \overline{B}_{1/k}, \\ v_k = \sup_{|x|=1/k} u & \text{on } \partial B_{1/k}, \\ v_k = \sup_{|x|=1/2} u & \text{on } \partial B_{1/2}. \end{cases}$$

Observe that u is a subsolution, while $c\Phi_{m,\beta}$ is a supersolution of the above problem for suitable $c > 0$. By the maximum principle, we find that $k \longmapsto v_k$ is increasing and

$$c\Phi_{m,\beta} \geq v_k \geq u \qquad \text{in } B_{1/2} \setminus B_{1/k}, \tag{3.39}$$

for some constant $c > 0$. Thus, there exists $v(x) = \lim_{k \to \infty} v_k(x)$ for all $x \in \overline{B}_{1/2} \setminus \{0\}$ and $v \in C^1(B_{1/2} \setminus \{0\})$ satisfies

$$\operatorname{div}\big(|x|^{-\beta}|\nabla v|^{m-2}\nabla v\big) = C|x|^{N-\alpha-\sigma p}v^q \quad \text{in } B_1 \setminus \{0\}. \tag{3.40}$$

Also v is radial (since v_k is radial) and from (3.39) we find

$$c\Phi_{m,\beta} \geq v \geq u \quad \text{in } B_{1/2} \setminus \{0\}. \tag{3.41}$$

Using this inequality, it is easy to see that v satisfies the conditions of Proposition B.3 with

$$a(x) = |x|^{N-\alpha-\sigma p}v(x)^{q-m+1} \leq c|x|^{N-\alpha-\sigma(p+q-m+1)} = c|x|^{-m-\beta}.$$

Thus, by the strong Harnack inequality (B.4), (3.38) and (3.41), we find

$$v(x) \geq c \sup_{|y|=|x|} v(y) \geq C|x|^{-\sigma} \quad \text{for all } x \in B_{1/4} \setminus \{0\}.$$

From (3.40) and the above estimate, we find

$$\big(r^{N-1-\beta}|v'(r)|^{m-2}v'(r)\big)' = Cr^{2N-\alpha-1-\sigma p}v^q \geq Cr^{2N-\alpha-1-\sigma(p+q)} \quad \text{for all } 0 < r < 1/4.$$

Since

$$\sigma = \frac{N-m-\beta}{m-1} = \frac{N+m-\alpha+\beta}{p+q-m+1},$$

the above estimate reads

$$\big(r^{N-1-\beta}|v'(r)|^{m-2}v'(r)\big)' \geq Cr^{-1} \quad \text{for all } 0 < r < 1/4.$$

We now fix $\bar{r} \in (0, 1/4)$ and integrate in the above inequality over $[r, \bar{r}]$. We obtain

$$-r^{N-1-\beta}|v'(r)|^{m-2}v'(r) \geq -\bar{r}^{N-1-\beta}|v'(\bar{r})|^{m-2}v'(\bar{r}) + C\ln\frac{\bar{r}}{r} \quad \text{for all } 0 < r < \bar{r} < 1/4.$$

From here we have

$$\lim_{r\to 0^+} r^{N-1-\beta}|v'(r)|^{m-2}v'(r) = -\infty,$$

so that

$$\lim_{r\to 0^+} \frac{v'(r)}{r^{-\frac{N-1-\beta}{m-1}}} = -\infty.$$

By l'Hopital's theorem, it follows that

$$\lim_{r \to 0^+} \frac{v(r)}{\Phi_{m,\beta}(r)} = \infty,$$

which contradicts (3.41) and proves our claim. Hence, $\sigma > \frac{N-m-\beta}{m-1}$ which yields (3.29)$_2$. Also, (3.29)$_3$ follows from (3.29)$_2$ and the fact that $p, q > m - 1$.

To derive the first inequality in (3.29), we combine the weak Harnack inequality and Proposition 3.5 with the regularity condition $u \in L^p(B_1)$. We find

$$\infty > \int_{B_{1/2}} u^p \geq \sum_{k=1}^{\infty} \int_{2^{-1-3k} < |y| < 2^{2-3k}} u^p(y) dy$$

$$\geq \frac{C}{2^N} \sum_{k=1}^{\infty} 2^{-3kN} \left(\sup_{\frac{5}{2} \cdot 2^{-3k} < |y| < \frac{7}{2} \cdot 2^{-3k}} u(y) \right)^p \quad \text{(by (B.6) with } R = 2^{-1-3k}, a = 7, b = 5, c = 1/2)$$

$$\geq \frac{C}{2^N} \sum_{k=1}^{\infty} 2^{-3kN} \left(\sup_{|y| = 3 \cdot 2^{-3k}} u(y) \right)^p$$

$$\geq \frac{C}{2^N \cdot 3^{\frac{N-m-\beta}{m-1}p}} \sum_{k=1}^{\infty} \frac{1}{(8^{N-\frac{N-m-\beta}{m-1}p})^k}. \quad \text{(by Proposition 3.5)}$$

This implies $N - \frac{N-m-\beta}{m-1} p > 0$ which establishes the first inequality in (3.29) for p. If $q \geq \frac{N(m-1)}{N-m-\beta}$, by Theorem 3.1 we deduce $u \in L^\infty(B_1)$, which is not possible since u is singular. Hence, $q < \frac{N(m-1)}{N-m-\beta}$.

Conversely, assume that (3.29) holds. We construct a singular radially symmetric solution u of (3.1) in the form $u(x) = \kappa|x|^{-\gamma}$, with $\kappa, \gamma > 0$ to be determined.

Case 1: $\sigma p > N - \alpha$.

Let

$$\max\left\{ \frac{N-m-\beta}{m-1}, \frac{N-\alpha}{p} \right\} < \gamma < \min\left\{ \sigma, \frac{N}{p} \right\}. \tag{3.42}$$

Note that this choice of γ is possible, thanks to (3.29)$_1$ and to our assumption $\sigma > \frac{N-m-\beta}{m-1}$. Also, $u(x) = \kappa|x|^{-\gamma}$ satisfies (3.36), where now the positivity of A follows from the lower bound of γ.

By Lemma 3.8(i) with $\theta = 0$, $a = \alpha$, $b = p\gamma$ so that $a + b > N$ and $p\gamma > N - \alpha$, we find

$$(|x|^{-\alpha} * u^p)u^q(x) \leq C\kappa^p |x|^{N-\alpha-p\gamma} u^q(x) \leq C\kappa^{p+q} |x|^{N-\alpha-(p+q)\gamma} \quad \text{in } B_1 \setminus \{0\}. \tag{3.43}$$

Using (3.36), (3.43) and the fact that $p + q > m - 1$ together with $\gamma < \sigma$, we may take $\kappa \in (0, 1)$ small enough such that

$$\text{div}\left(|x|^{-\beta}|\nabla u|^{m-2}\nabla u\right) = C_1\kappa^{m-1}|x|^{-(m-1)\gamma-m-\beta}$$

$$\geq C_2\kappa^{p+q}|x|^{N-\alpha-(p+q)\gamma}$$

$$\geq (|x|^{-\alpha}*u^p)u^q(x) \quad \text{in } B_1\setminus\{0\}.$$

This shows that $u(x) = \kappa|x|^{-\gamma}$ is a positive singular solution of (3.1) in $B_1\setminus\{0\}$.

Case 2: $\sigma p \leq N - \alpha$.

Let us observe first that this condition is equivalent to

$$\frac{m+\beta}{q-m+1} \leq \sigma \leq \frac{N-\alpha}{p}.$$

Also, by $(3.29)_1$ we have

$$\frac{N-m-\beta}{m-1} < \frac{m+\beta}{q-m+1}.$$

Let

$$\frac{N-m-\beta}{m-1} < \gamma < \frac{m+\beta}{q-m+1} \leq \sigma \leq \frac{N-\alpha}{p}.$$

Letting now $u(x) = \kappa|x|^{-\gamma}$, we have that u satisfies (3.36), where here $A > 0$ by the lower bound of γ. Also, by Lemma 3.8(iii) with $\theta = 0$, $a = \alpha$, $b = p\gamma$ so that $a + b < N$ and $p\gamma < N - \alpha$, we have

$$(|x|^{-\alpha}*u^p)u^q(x) \leq C\kappa^p u^q(x) \leq C\kappa^{p+q}|x|^{-q\gamma} \quad \text{in } B_1\setminus\{0\}. \tag{3.44}$$

Combining (3.36) and (3.44) in the same way as we did in Case 1, we derive that $u(x) = \kappa|x|^{-\gamma}$ is a singular solution of (3.1).

$\square$

3.5 A Double Inequality

In this section, we study the existence and the asymptotic behaviour of singular solutions to the double inequality (3.2). Let $\sigma > 0$ be given by (3.26).

Theorem 3.10 *Assume $m > 1$, $p, q > m - 1$, $\alpha \in (0, N)$, $\beta \in \mathbb{R}$, $N \geq 1$ and $\sigma p < N$.*

(i) *(Existence)*

  (i1) *If $N > m + \beta$, then there exists $a \geq b > 0$ and a singular solution of (3.2) if and only if (3.29) holds;*
  (i2) *If $N \leq m + \beta$, then (3.2) has always singular solutions for some $a \geq b > 0$.*

(ii) *(Asymptotic behavior) Assume $N \geq m + \beta$ and*

$$\begin{cases} m - 1 < q < \dfrac{N - (\sigma p + \alpha - N)^+}{N - m - \beta}(m - 1) & \text{if } N > m + \beta, \\ m - 1 < q < \infty & \text{if } N = m + \beta. \end{cases}$$
$$(3.45)$$

(ii1) *If* $\sigma p > N - \alpha$, *then any singular solution of* (3.2) *satisfies*

$$\text{either} \quad u(x) \asymp \Phi_{m,\beta}(x) \quad or \quad u(x) \asymp |x|^{-\sigma}. \qquad (3.46)$$

(ii2) *If* $\sigma p < N - \alpha$, *then any singular solution of* (3.2) *satisfies*

$$\text{either} \quad u(x) \asymp \Phi_{m,\beta}(x) \quad or \quad u(x) \asymp |x|^{-\frac{m+\beta}{q-m+1}}. \qquad (3.47)$$

Theorem 3.10(ii) states that any singular solution u of (3.2) either behaves like the fundamental solution $\Phi_{m,\beta}(x)$ in a neighbourhood of the origin or has a stronger singularity precisely given by $(3.46)_2$–$(3.47)_2$. In particular, the asymptotic behaviour in Theorem 3.10(ii) applies to singular solutions of the equation

$$\operatorname{div}\left(|x|^{-\beta}|\nabla u|^{m-2}\nabla u\right) = (|x|^{-\alpha} * u^p)u^q \quad \text{in} \quad B_1 \setminus \{0\}.$$

The approach relies on the following result.

Lemma 3.11 *Let* $m > 1$, $N \geq m + \beta > \theta$ *and*

$$\begin{cases} m - 1 < q < \dfrac{(N - \theta)(m - 1)}{N - m - \beta} & \text{if } N > m + \beta, \\ m - 1 < q < \infty & \text{if } N = m + \beta. \end{cases} \qquad (3.48)$$

Let $u \in W^{1,m}(B_1 \setminus \{0\}) \cap C(B_1 \setminus \{0\})$, $u \geq 0$, *be a singular solution of*

$$\operatorname{div}(|x|^{-\beta}|\nabla w|^{m-2}\nabla w) = |x|^{-\theta}w^q \quad \text{in } B_1 \setminus \{0\}. \qquad (3.49)$$

Then, either $w \asymp \Phi_{m,\beta}(x)$ *or* $w \asymp |x|^{-\frac{m+\beta-\theta}{q-m+1}}$.

We will not prove this result here; it relies on the study of radial solutions to (3.49) which satisfy

$$(r^{N-1-\beta}|w_r|^{m-2}w_r)_r = r^{N-1-\theta}w \quad \text{for } 0 < r = |x| < 1. \qquad (3.50)$$

Using the change of variable $y(s) = w(r)$, $s = \Phi_{m,\beta}(r)$, we are led to

$$(|y_s|^{m-2}y_s)_r = C(N)s^{N-\theta+(N-\beta-m)/(m-1)}y^q \quad \text{for } s > 1 \text{ large}.$$

We refer the reader to [MNU02] for the study of the asymptotic profile of y and thus of the solution w to (3.50).

Proof of Theorem 3.10

(i) Any singular solution u of (3.2) fulfills in particular (3.1). Thus, by Theorem 3.7 conditions (3.29) must hold if $N > m + \beta$.

Conversely, assume now that either $N \le m + \beta$ or $N > m + \beta$ and (3.29) holds. Let σ be defined by (3.26) and $\tau = \frac{1}{p+q-m+1} > 0$.

We claim that

$$
u(x) = \begin{cases}
|x|^{-\sigma} & \text{if } \sigma p > N - \alpha, \\[2mm]
|x|^{-\sigma} \left(\log \frac{5}{|x|} \right)^{-\tau} & \text{if } \sigma p = N - \alpha, \\[2mm]
|x|^{-\frac{m+\beta}{q-m+1}} & \text{if } \sigma p < N - \alpha,
\end{cases}
$$

is a solution of (3.2). A straightforward calculation using Remark 3.9 yields

$$
\operatorname{div}\!\left(|x|^{-\beta}|\nabla u|^{m-2}\nabla u\right) \asymp \begin{cases}
|x|^{-\sigma(m-1)-m-\beta} & \text{if } \sigma p > \alpha, \\[2mm]
|x|^{-\sigma(m-1)-m-\beta} \left(\log \frac{5}{|x|} \right)^{-(m-1)\tau} & \text{if } \sigma p = \alpha, \\[2mm]
|x|^{-\frac{q(m+\beta)}{q-m+1}} & \text{if } \sigma p < \alpha.
\end{cases}
$$

To see this, we first note that $(3.29)_2$ implies

$$
\sigma > \frac{N - m - \beta}{m - 1}. \tag{3.51}
$$

Thus, the coefficient A defined in (3.34) (in which $\gamma = \sigma$) satisfies $A > 0$.

Also, by Lemma 3.8(i)–(iii) (we use $\theta = \tau p \in (0,1)$ if $\sigma p = N - \alpha$), we have

$$
(|x|^{-\alpha} * u^p)u^q(x) \asymp \begin{cases}
|x|^{N-\alpha-\sigma(p+q)} & \text{if } \sigma p > N - \alpha, \\[2mm]
|x|^{-q\sigma} \left(\log \frac{5}{|x|} \right)^{1-\tau(p+q)} & \text{if } \sigma p = N - \alpha, \\[2mm]
|x|^{-\frac{q(m+\beta)}{q-m+1}} & \text{if } \sigma p < N - \alpha.
\end{cases}
$$

From the above estimates, we have

$$
\operatorname{div}\!\left(|x|^{-\beta}|\nabla u|^{m-2}\nabla u\right) \asymp (|x|^{-\alpha} * u^p)u^q
$$

and thus, for suitable constants $a \ge b > 0$, we have that u satisfies (3.2).

(ii) Let u be a singular solution of (3.2). We divide our argument into two steps.

Step 1: u satisfies the strong Harnack inequality (B.4).

Note first that u satisfies the inequality

$$\mathrm{div}\big(|x|^{-\beta}|\nabla u|^{m-2}\nabla u\big) \geq cu^q \quad \text{in } B_1 \setminus \{0\},$$

where $c = 2^{-\alpha} \int_{B_1} u^p dx > 0$. Applying Proposition 3.3 with $\theta = 0$, we find

$$u(x) \leq C|x|^{-\frac{m+\beta}{q-m+1}} \quad \text{in } B_1 \setminus \{0\}. \tag{3.52}$$

Using the above estimate (if $\sigma p < N - \alpha$) and (3.25) (if $\sigma p > N - \alpha$), from Lemma 3.8(i),(iii) we obtain

$$(|x|^{-\alpha} * u^p)(x) \leq C\varphi(x) \quad \text{in } B_1 \setminus \{0\}, \tag{3.53}$$

where

$$\varphi(x) = |x|^{-(\sigma p + \alpha - N)^+}. \tag{3.54}$$

(we take $(\sigma p + \alpha - N)^+ = 0$ if $\sigma p + \alpha - N \leq 0$). Now, (3.53) together with (3.52) (if $\sigma p < N - \alpha$) and (3.25) (if $\sigma p > N - \alpha$) imply

$$(|x|^{-\alpha} * u^p)u^{q-m+1} \leq C|x|^{-m-\beta} \quad \text{in } B_1 \setminus \{0\}.$$

We are exactly in the frame of Proposition B.3 which yields (B.4).

Step 2: Proof of (3.46)–(3.47).

Our analysis is split into two cases.

Case 1: Suppose

$$\limsup_{x \to 0} \frac{u(x)}{\Phi_{m,\beta}(x)} < \infty. \tag{3.55}$$

Let $c > 0$ be such that $u(x) \leq c\Phi_{m,\beta}(x)$ in $\overline{B}_1 \setminus \{0\}$. By Lemma 3.8 we have

$$|x|^{-\alpha} * u^p \leq |x|^{-\alpha} * (c\Phi_{m,\beta})^p \leq C|x|^{-\theta} \quad \text{in } B_1 \setminus \{0\}, \tag{3.56}$$

where

$$\theta = \begin{cases} p\dfrac{N-m-\beta}{m-1} + \alpha - N & \text{if } p\dfrac{N-m-\beta}{m-1} > N - \alpha, \\[2mm] \tau & \text{if } p\dfrac{N-m-\beta}{m-1} = N - \alpha, \\[2mm] 0 & \text{if } p\dfrac{N-m-\beta}{m-1} < N - \alpha, \end{cases} \tag{3.57}$$

and $\tau > 0$ is chosen small enough such that[1]

$$q < \frac{N - (\sigma p + \alpha - N)^+ - \tau}{N - m - \beta}(m - 1).$$

Also, by the definition of θ in (3.57) and (3.29), we have $0 \le \theta < m + \beta$; this latter condition is required in the statement of Lemma 3.11. Indeed, this is easy to check if $p\frac{N-m-\beta}{m-1} \le N - \alpha$. If $p\frac{N-m-\beta}{m-1} > N - \alpha$, then we observe that from $(3.29)_2$ and $q > m - 1$, we find

$$p < \frac{N + m - \alpha + \beta}{N - m - \beta}(m - 1),$$

and then

$$\theta = p\frac{N - m - \beta}{m - 1} + \alpha - N < m + \beta.$$

Since u is a singular solution of (3.2), there exists a decreasing sequence $\{r_k\} \subset (0, 1)$, $r_k \to 0$ (as $k \to \infty$) such that

$$\sup_{|x|=r_k} u(x) \to \infty \quad \text{as } k \to \infty.$$

Using the strong Harnack inequality (B.4), we also have

$$\inf_{|x|=r_k} u(x) \to \infty \quad \text{as } k \to \infty. \tag{3.58}$$

For any $k \ge 1$, let $w_k \in C^1(B_1 \setminus B_{r_k})$ be a radial function such that

$$\begin{cases} \operatorname{div}\left(|x|^{-\beta}|\nabla w_k|^{m-2}\nabla w_k\right) = C|x|^{-\theta}w_k^q & \text{in } B_1 \setminus \overline{B}_{r_k}, \\[2mm] w_k = \inf_{|x|=r_k} u & \text{on } \partial B_{r_k}, \\[2mm] w_k = \inf_{|x|=1} u & \text{on } \partial B_1. \end{cases}$$

Since u satisfies (3.56), by the comparison principle in Proposition 2.6, we find that $k \longmapsto w_k$ is increasing and $u \ge w_k$ in $B_1 \setminus B_{r_k}$. Thus, there exists $w(x) = \lim_{k \to \infty} w_k(x)$ for all $x \in \overline{B}_1 \setminus \{0\}$ and $w \in C^1(B_1 \setminus \{0\})$ satisfies

$$\operatorname{div}\left(|x|^{-\beta}|\nabla w|^{m-2}\nabla w\right) = C|x|^{-\theta}w^q \quad \text{in } B_1 \setminus \{0\}.$$

---

[1] We choose τ with the above conditions just to avoid the log terms that appear in estimating the convolution integrals.

Also w is singular since by (3.58) we have

$$\sup_{|x|=r_k} w \geq \sup_{|x|=r_k} w_k = \inf_{|x|=r_k} u \to \infty \quad \text{as } k \to \infty.$$

Thus, by (3.55) and Lemma 3.11, we deduce $u \asymp \Phi_{m,\beta}$.

Case 2:   Suppose

$$\limsup_{x \to 0} \frac{u(x)}{\Phi_{m,\beta}(x)} = \infty.$$

Hence, one may find a decreasing sequence $\{r_k\} \subset (0, 1)$, $r_k \to 0$ (as $k \to \infty$) such that

$$\sup_{|x|=r_k} \frac{u(x)}{\Phi_{m,\beta}(x)} \to \infty \quad \text{as } k \to \infty.$$

Using the strong Harnack inequality (B.4) for u and the fact that $\Phi_{m,\beta}(x) = \Phi_{m,\beta}(|x|)$, one has

$$\inf_{|x|=r_k} \frac{u(x)}{\Phi_{m,\beta}(x)} \to \infty \quad \text{as } k \to \infty. \tag{3.59}$$

Recall that u satisfies (3.53)–(3.54). For any $k \geq 1$, let $w_k \in C^1(B_1 \setminus B_{r_k})$ be a radial function such that

$$\begin{cases} \operatorname{div}\left(|x|^{-\beta}|\nabla w_k|^{m-2}\nabla w_k\right) = C\varphi(x)w_k^q & \text{in } B_1 \setminus \overline{B}_{r_k}, \\ w_k = \inf_{|x|=r_k} u & \text{on } \partial B_{r_k}, \\ w_k = \inf_{|x|=1} u & \text{on } \partial B_1, \end{cases}$$

where φ is defined in (3.53). By the comparison principle in Proposition 2.6, we find that $k \longmapsto w_k$ is increasing and $u \geq w_k$ in $B_1 \setminus B_{r_k}$. Thus, there exists $w(x) = \lim_{k \to \infty} w_k(x)$ for all $x \in \overline{B}_1 \setminus \{0\}$ and

$$\operatorname{div}\left(|x|^{-\beta}|\nabla w|^{m-2}\nabla w\right) = C\varphi(x)w^q \quad \text{in } B_1 \setminus \{0\}.$$

In particular, w satisfies (3.49) with $\theta = (\sigma p + \alpha - N)^+ < m + \beta$ since $q > m - 1$, and

$$u \geq w \geq w_k \quad \text{in } B_1 \setminus B_{r_k}. \tag{3.60}$$

Using the above estimates and (3.59), it follows that

$$\limsup_{x \to 0} \frac{u(x)}{\Phi_{m,\beta}(x)} \geq \lim_{k \to 0} \sup_{|x|=r_k} \frac{w(x)}{\Phi_{m,\beta}(x)} \geq \lim_{k \to 0} \sup_{|x|=r_k} \frac{w_k(x)}{\Phi_{m,\beta}(x)} = \infty.$$

By Lemma 3.11 it follows that

$$w \asymp |x|^{-\dfrac{m + \beta - (\sigma p + \alpha - N)^+}{q - m + 1}}.$$

This fact combined with (3.52) (if $\sigma p < N - \alpha$) and (3.25) (if $\sigma p > N - \alpha$) implies the estimates in Theorem 3.10(ii).

$\square$

3.6 Conclusions and Further Remarks

In this chapter, we studied the positive solutions for a class of quasilinear inequalities in the punctured unit ball featuring nonlinear convolution terms. We obtained an L_{loc}^∞ regularity result and then the focus was turned to the existence and asymptotic of singular solutions. Crucial to our approach is the Keller-Osserman type estimate, the weak and strong Harnack inequalities and the properties of the solution set to

$$\text{div}(|x|^{-\beta}|\nabla u|^{m-2}\nabla u) = |x|^{-\theta}u^q \quad \text{in } B_1 \setminus \{0\}. \tag{3.61}$$

Equations of type (3.61) and their semilinear counterpart have been studied since the early 1980s by H. Brézis, A. Friedman, L. Véron and J.L. Vàzquez: [BV81], [FV86], [Ver81] and [VV80]. More recent extensions appear in [BCCT13], [SYW16]. The nonlocal inequalities (3.1) and (3.2) were studied in [FG20]. According to Theorem 3.10, any positive singular solution of

$$\text{div}(|x|^{-\beta}|\nabla u|^{m-2}\nabla u) = (|x|^{-\alpha} * u^p)u^q \quad \text{in } B_1 \setminus \{0\} \tag{3.62}$$

satisfies either (3.46) or (3.47). An open problem in this direction is that any positive singular solution u of (3.62) follows one of the asymptotic behaviours below:

- either $\dfrac{u(x)}{\Phi_{m,\beta}(x)} \to A > 0$ as $|x| \to 0$;

- or $\dfrac{u(x)}{|x|^{-\sigma}} \to B > 0$ as $|x| \to 0$;

- or $\dfrac{u(x)}{|x|^{-\frac{m+\beta}{q-m+1}}} \to C > 0$ as $|x| \to 0$,

where $A, B, C > 0$ are constants depending on N, m, α, β, p and q.

Chapter 4
Polyharmonic Inequalities with Convolution Terms

4.1 Introduction

In this chapter, we are concerned with the higher-order elliptic inequality

$$\pm \Delta^m u \geq \left(K(|x|) * u^p \right) u^q \quad \text{in } \mathbb{R}^N, \tag{4.1}$$

and its corresponding systems of inequalities, where $N, m \geq 1$ are integers, $p, q > 0$ and Δ^m denotes the m-polyharmonic operator, that is,

$$\Delta^m u = \Delta(\Delta^{m-1} u).$$

The function K satisfies

$$\begin{cases} K > 0 \quad \text{and} \quad K(|x|) \in C(\mathbb{R}^N \setminus \{0\}) \cap L^1_{loc}(\mathbb{R}^N); \\ K(r) \text{ is non-increasing on } (0, \infty); \\ \lim_{r \to \infty} r^N K(r) = \infty. \end{cases} \tag{4.2}$$

Typical examples of functions K include:

$$K(r) = r^{-\alpha}, \quad 0 < \alpha < N \quad \text{or} \quad K(r) = r^{-N} \log^{-\beta}\left(1 + \frac{1}{r}\right), \quad 1 < \beta \leq N.$$

As in the previous sections, the operator $*$ in (4.1) is the standard convolution; see (1.3). By a non-negative solution u of (4.1) we understand a function $u \subset C^{2m}(\mathbb{R}^N)$, $u \geq 0$, such that

© The Author(s), under exclusive license to Springer Nature Switzerland AG 2022
M. Ghergu, *Partial Differential Inequalities with Nonlinear Convolution Terms*,
SpringerBriefs in Mathematics, https://doi.org/10.1007/978-3-031-21856-9_4

$$\int_{|y|>1} K\left(\frac{|y|}{2}\right) u^p(y)dy < \infty \tag{4.3}$$

and u satisfies (4.1) pointwise.

Note that condition (4.3) is weaker than the condition $u \in L^p(\mathbb{R}^N)$ and that (4.3) is (almost) necessary and sufficient in order to ensure that the convolution term $K(|x|) * u^p$ is finite for all $x \in \mathbb{R}^N$. Indeed, for any $x \in \mathbb{R}^N$ we have

$$K(|x|) * u^p = \int_{|y|\leq 2|x|} u^p(y)K(|x-y|)dy + \int_{|y|>2|x|} K(|x-y|)u^p(y)dy$$

$$\leq \left(\max_{|z|\leq 2|x|} u(z)\right)^p \int_{B_{3|x|}} K(|y|)dy + \int_{|y|>2|x|} K\left(\frac{|y|}{2}\right) u^p(y)dy < \infty,$$

by the fact that $K(|x|) \in L^1_{loc}(\mathbb{R}^N)$ is nonincreasing and (4.3).

In the sequel, we prefer to separate the analysis of (4.1) into two distinct inequalities as follows:

$$-(-\Delta)^m u \geq \left(K(|x|) * u^p\right)u^q \quad \text{in } \mathbb{R}^N,$$

and

$$(-\Delta)^m u \geq \left(K(|x|) * u^p\right)u^q \quad \text{in } \mathbb{R}^N.$$

4.2 An Integral Representation Formulae

A key tool in our approach is the fact that for positive Radon measures μ on $\mathbb{R}^N$, solutions to

$$(-\Delta)^m u = \mu \quad \text{in } \mathscr{D}'(\mathbb{R}^N), \tag{4.4}$$

allow, under some conditions, an integral form representation. Solutions of (4.4) are understood in the sense of distributions, that is,

$$\int_{\mathbb{R}^N} u(-\Delta)^m \varphi \, dx = \int_{\mathbb{R}^N} \varphi(x)d\mu(x) \quad \text{for all } \varphi \in C^\infty_c(\mathbb{R}^N). \tag{4.5}$$

More precisely, we have:

Theorem 4.1 *Let $m \geq 1$ be an integer and $N > 2m$. Suppose μ is a positive Radon measure on $\mathbb{R}^N$ and $\ell \in \mathbb{R}$. The following statements are equivalent:*

(i) *$u \in L^1_{loc}(\mathbb{R}^N)$ is a distributional solution of (4.4) and for a.e. $x \in \mathbb{R}^N$, u satisfies the ring condition we have*

$$\liminf_{R \to \infty} \frac{1}{R^N} \int_{R \le |y-x| \le 2R} |u(y) - \ell| dy = 0. \tag{4.6}$$

(ii) $u \in L^1_{loc}(\mathbb{R}^N)$ *is a distributional solution of (4.4), essinf $u = \ell$ and u is weakly superharmonic in the sense that*

$$\int_{\mathbb{R}^N} u(-\Delta)^j \varphi \ge 0 \quad \text{for all } 1 \le j \le m, \varphi \in C_0^\infty(\mathbb{R}^N), \varphi \ge 0.$$

(iii) $u \in L^1_{loc}(\mathbb{R}^N)$ *and there exists $c = c(N, m) > 0$ such that*

$$u(x) = \ell + c \int_{\mathbb{R}^N} \frac{d\mu(y)}{|x - y|^{N-2m}} \quad \text{for a.e. } x \in \mathbb{R}^N. \tag{4.7}$$

Proof We shall prove that the following sequence of implications hold:

$$\text{(i)}\Longrightarrow\text{(iii)}\Longrightarrow\text{(ii)\&(i)} \quad \text{and} \quad \text{(ii)}\Longrightarrow\text{(i)}.$$

Replacing u by $u - \ell$ in the following, we may assume $\ell = 0$.
(i)$\Longrightarrow$(iii) Let

$$g(x) = (1 + |x|^2)^{-\frac{N+2m}{2}}$$

so that $g \in L^1(\mathbb{R}^N)$ and

$$\|g\|_{L^1(\mathbb{R}^N)} = \frac{\pi^{N/2}(m - 1)!}{\Gamma\left(\frac{N+2m}{2}\right)},$$

where Γ is the standard Gamma function. For $\varepsilon > 0$ set $g_\varepsilon(x) = \varepsilon^{-N} g(x/\varepsilon)$. Then,

$$g_\varepsilon * u \to \|g\|_{L^1(\mathbb{R}^N)} u \quad \text{in } L^1(\mathbb{R}^N) \quad \text{as } \varepsilon \to 0.$$

Passing to a subsequence which we still denote $\{g_\varepsilon\}$, we have

$$g_\varepsilon * u \to \|g\|_{L^1(\mathbb{R}^N)} u \quad \text{a.e. in } \mathbb{R}^N \quad \text{as } \varepsilon \to 0. \tag{4.8}$$

We claim that the representation formula (4.7) holds for all points $x \in \mathbb{R}^N$ for which (4.6) and the pointwise convergence (4.8) hold. Without loosing the generality, we may assume that (4.6) and (4.8) hold for $x = 0$. Hence,

$$(g_\varepsilon * u)(0) \to \|g\|_{L^1(\mathbb{R}^N)} u(0) = \frac{\pi^{N/2}(m - 1)!}{\Gamma\left(\frac{N+2m}{2}\right)} u(0) \quad \text{as } \varepsilon \to 0. \tag{4.9}$$

We shall prove that the representation formula (4.7) holds for $x = 0$.

Let $\phi \in C_c^\infty(\mathbb{R})$ be a standard cutoff function which satisfies

$$0 \leq \phi \leq 1, \quad \phi \equiv 1 \text{ on } [-1, 1] \quad \text{and} \quad \mathrm{supp}\,\phi = [-2, 2]. \tag{4.10}$$

For $R \geq 1$ let $\phi_R(x) = \phi(|x|/R)$. From (4.6) we can find an increasing sequence $\{R_j\}$ that tends to infinity such that

$$\lim_{j \to \infty} \frac{1}{R^N} \int_{R_j \leq |y| \leq 2R_j} |u(y)|\,dy = 0. \tag{4.11}$$

From (4.9) for all $j \geq 1$, we also have that

$$(g_\varepsilon * u\phi_{R_j})(0) \to \|g\|_{L^1(\mathbb{R}^N)} u(0) = \frac{\pi^{N/2}(m-1)!}{\Gamma\left(\frac{N+2m}{2}\right)} u(0) \quad \text{as } \varepsilon \to 0. \tag{4.12}$$

For $\varepsilon, \rho > 0$ define

$$\theta_{\varepsilon,\rho}(x) = (\varepsilon^2 + |x|^2)^{-\frac{N-2m}{2}} \phi\left(\frac{|x|}{\rho}\right).$$

$\square$

Lemma 4.2 *For any* $1 \leq m < N/2$, *we have*

$$(-\Delta)^m (\varepsilon^2 + |x|^2)^{-\frac{N-2m}{2}} = 2^{2m} \frac{\Gamma\left(\frac{N+2m}{2}\right)}{\Gamma\left(\frac{N-2m}{2}\right)} \gamma_\varepsilon(x).$$

Proof For any $q > 0$, define $U_q = (\varepsilon^2 + |x|^2)^{-q/2}$. A straightforward computation yields

$$-\Delta U_q = q(N - q - 2)U_{q+2} + \varepsilon^2 q(q + 2)U_{q+4}.$$

An induction argument over $m \geq 1$ allows us to conclude. $\square$

Let Φ be the Riesz kernel defined as

$$\Phi(x) = \frac{\Gamma\left(\frac{N-2m}{2}\right)}{2^{2m} \pi^{N/2}(m-1)!} |x|^{2m-N}, \quad x \in \mathbb{R}^N. \tag{4.13}$$

It turns out that for $N > 2m$, the function Φ is the fundamental solution of the operator $(-\Delta)^m$, that is,

$$(-\Delta)^m \Phi = \delta_0 \quad \text{in } \mathscr{D}'(\mathbb{R}^N),$$

where δ_0 is the Dirac delta measure concentrated at the origin in $\mathbb{R}^N$.

Choose in (4.5) the test function $\varphi = \theta_{\varepsilon,\rho}\phi_{R_j}$. Hence,

$$\int_{\mathbb{R}^N} \theta_{\varepsilon,\rho}\phi_{R_j} d\mu = \int_{\mathbb{R}^N} u(-\Delta)^m \big(\theta_{\varepsilon,\rho}\phi_{R_j}\big)dx = I_1 + I_2, \tag{4.14}$$

where

$$I_1 = \int_{\mathbb{R}^N} u\phi_{R_j}(-\Delta)^m\theta_{\varepsilon,\rho}dx,$$

$$I_2 = \int_{\mathbb{R}^N} u\Big\{(-\Delta)^m \big(\theta_{\varepsilon,\rho}\phi_{R_j}\big) - \phi_{R_j}(-\Delta)^m\theta_{\varepsilon,\rho}\Big\}dx.$$

Taking into account that on B_{R_j} for $\rho > 2R_j$, one has $\theta_{\varepsilon,\rho}(x) = (\varepsilon^2 + |x|^2)^{-\frac{N-2m}{2}}$, and letting $\rho \to \infty$ in the above equality, we find

$$I_1 \longrightarrow J_1 := \int_{\mathbb{R}^N} u\phi_{R_j}(-\Delta)^m \big((\varepsilon^2 + |x|^2)^{-\frac{N-2m}{2}}\big)dx.$$

Thus, by Lemma 4.2 we deduce

$$J_1 = \int_{\mathbb{R}^N} u\phi_{R_j}(-\Delta)^m \big((\varepsilon^2 + |x|^2)^{-\frac{N-2m}{2}}\big)dx$$

$$= 2^{2m}\frac{\Gamma\big(\frac{N+2m}{2}\big)}{\Gamma\big(\frac{N-2m}{2}\big)} \int_{\mathbb{R}^N} u\phi_{R_j}g_\varepsilon(x)dx$$

$$= 2^{2m}\frac{\Gamma\big(\frac{N+2m}{2}\big)}{\Gamma\big(\frac{N-2m}{2}\big)} \big(g_\varepsilon * u\phi_{R_j}\big)(0).$$

Next, using the convergence (4.12), we find

$$J_1 \to \frac{2^{2m}\pi^{N/2}(m-1)!}{\Gamma\big(\frac{N-2m}{2}\big)}u(0) \quad \text{as } \varepsilon \to 0, \quad \text{for all } j \geq 1. \tag{4.15}$$

To estimate I_2, we first observe that letting $\rho \to \infty$ and then $\varepsilon \to 0$, we have

$$I_2 \to J_2 := \frac{2^{2m}\pi^{N/2}(m-1)!}{\Gamma\big(\frac{N-2m}{2}\big)} \int_{R_j<|x|<2R_j} uL_j(x)dx,$$

where $L_j = (-\Delta)^m \big(\Phi\phi_{R_j}\big) - \phi_{R_j}(-\Delta)^m\Phi$. Thus, L_j is a linear combination of quantities of the form $D^\alpha\Phi(x)D^\beta\phi_{R_j}(x)$ where α, β are multi-indices such that $|\beta| > 0$ and $|\alpha| + |\beta| = 2m$. We have

$$\big|D^\alpha\Phi(x)\big| \leq cR_j^{2m-|\alpha|-N}, \quad \big|D^\beta\phi_{R_j}(x)\big| \leq cR_j^{-|\beta|} \quad \text{for all } R_j < |x| < 2R_j.$$

Thus,

$$|L_j(x)| \le \frac{c}{R_j^N} \quad \text{for all } R_j < |x| < 2R_j.$$

Hence, by (4.11) we deduce

$$|J_2| \le C \int_{R_j < |x| < 2R_j} |u||L_j| dx \le \frac{c}{R^N} \int_{R_j < |x| < 2R_j} |u| dx \to 0 \quad \text{as} \quad j \to \infty.$$

$$(4.16)$$

Multiplying (4.14) by $\frac{\Gamma\left(\frac{N-2m}{2}\right)}{2^{2m}\pi^{N/2}(m-1)!}$ and letting $\rho \to \infty$ and then $\varepsilon \to 0$, by Beppo Levi monotone convergence theorem, we deduce

$$\int_{\mathbb{R}^N} \Phi \phi_{R_j} d\mu = u(0) + \frac{\Gamma\left(\frac{N-2m}{2}\right)}{2^{2m}\pi^{N/2}(m-1)!} J_2.$$

Letting $j \to \infty$ in the above estimate, using (4.16) and Beppo Levi monotone convergence theorem, we finally obtain

$$u(0) = \int_{\mathbb{R}^N} \Phi(y) d\mu(y),$$

which proves the representation (4.7).

(iii)$\Longrightarrow$(ii)&(i) For $1 \le j \le m$, let $\mathscr{R}_j$ be the Riesz kernel defined as

$$\mathscr{R}_j(x) = \frac{\Gamma\left(\frac{N-2j}{2}\right)}{2^{2j}\pi^{N/2}(j-1)!} |x|^{2j-N}, \quad x \in \mathbb{R}^N.$$

Then $\mathscr{R}_j$ is the fundamental solution of the operator $(-\Delta)^j$. Also, if Φ is the potential defined in (4.4), then $\Phi = \mathscr{R}_j * \mathscr{R}_{m-j}$ for all $1 \le j < m$.

Let $\tilde{\mu}$ defined for $x \in \mathbb{R}^N$, by

$$d\tilde{\mu}(x) = \left(\int_{\mathbb{R}^N} \mathscr{R}_{m-1}(x - y) d\mu(y) \right) dx$$

is a measure, and since

$$u(x) = (\Phi * \mu)(x) = (\mathscr{R}_1 * \mathscr{R}_{m-1} * \mu)(x) = (\mathscr{R}_1 * \tilde{\mu})(x)$$

is finite a.e., we deduce that $\tilde{\mu}$ is finite on compact sets, that is, $\tilde{\mu}$ is a Radon measure.

By Lemma A.2 it follows that u is superharmonic, $u \in L^1_{loc}(\mathbb{R}^N)$, u satisfies the ring condition (4.6) and $\text{essinf}_{\mathbb{R}^N} u = 0$. It remains to show that u is a distributional solution of (4.35).

Indeed, let $\varphi \in C_c^\infty(\mathbb{R}^N)$. By Fubini's theorem, we deduce

$$\int_{\mathbb{R}^N} u(-\Delta)^m \varphi dx = \int_{\mathbb{R}^N} (-\Delta)^m \varphi(x) \Big(\int_{\mathbb{R}^N} \Phi(x-y) d\mu(y) \Big) dx$$

$$= \int_{\mathbb{R}^N} \Big(\int_{\mathbb{R}^N} (-\Delta)^m \varphi(x) \Phi(x-y) dx \Big) d\mu(y) \qquad (4.17)$$

$$= \int_{\mathbb{R}^N} \varphi(y) d\mu(y),$$

so u is a distributional solution of (4.35).

Using $\Phi = \mathscr{R}_j * \mathscr{R}_{m-j}$, we find

$$u = \Phi * \mu = \mathscr{R}_j * \big(\mathscr{R}_{m-j} * \mu \big),$$

which yields $(-\Delta)^j u = \mathscr{R}_{m-j} * \mu$ in $\mathscr{D}'(\mathbb{R}^N)$ and in particular $(-\Delta)^j u \geq 0$ in $\mathscr{D}'(\mathbb{R}^N)$.

(ii)$\Longrightarrow$(i) This easily follows from Lemma A.1. Indeed, since u is superharmonic and $\text{essinf}_{\mathbb{R}^N} u = 0$, it follows from Lemma A.1 that

$$\lim_{R\to\infty} \fint_{B_R(x)} u(y)dy = \text{essinf}_{\mathbb{R}^N} u = 0.$$

Corollary 4.3 *Let $v \in L^1_{loc}(\mathbb{R}^N)$ be a distributional solution of*

$$(-\Delta)^m v = f \quad in \ \mathscr{D}'(\mathbb{R}^N), \ N > 2m,$$

where $f \in L^1_{loc}(\mathbb{R}^N)$, $f \geq 0$, $f \not\equiv 0$. Assume that

$$\int_{|y|>1} K\left(\frac{|y|}{2}\right) |v|^p(y)dy < \infty, \qquad (4.18)$$

where $p \geq 1$ and K is a function which satisfies (4.2). Then, $\text{essinf} \, v = 0$ and for some constant $c > 0$, we have

$$v(x) = c \int_{\mathbb{R}^N} \frac{f(y)}{|x-y|^{N-2m}} dy \quad for \ a.e. \ x \in \mathbb{R}^N. \qquad (4.19)$$

In particular $v > 0$ in $\mathbb{R}^N$ and

$$v(x) \geq C|x|^{2m-N} \quad in \ \mathbb{R}^N \setminus B_1, \qquad (4.20)$$

for some constant $C > 0$.

Proof We show that v satisfies condition (4.6) in Theorem 4.1 with $\ell = 0$. First, if $p > 1$ by Hölder's inequality and (4.18), for all $x \in \mathbb{R}^N$, we have

$$\int_{R\leq|y-x|\leq2R} |v(y)|dy \leq \Big(\int_{R\leq|y-x|\leq2R} K\Big(\frac{|y|}{2}\Big)|v|^p(y)dy \Big)^{\frac{1}{p}} \Big(\int_{R\leq|y-x|\leq2R} K^{-\frac{1}{p-1}}\Big(\frac{|y|}{2}\Big)dy \Big)^{1-\frac{1}{p}}$$

$$\leq C\Big(\int_{R\leq|y-x|\leq2R} K^{-\frac{1}{p-1}}\Big(\frac{|y|}{2}\Big)dy \Big)^{1-\frac{1}{p}} \tag{4.21}$$

$$= C\Big(\int_{R\leq|z|\leq2R} K^{-\frac{1}{p-1}}\Big(\frac{|z+x|}{2}\Big)dz \Big)^{1-\frac{1}{p}}.$$

Take $R > |x|$. By the property (4.2) on Ψ, we have

$$K^{-\frac{1}{p-1}}\Big(\frac{|z+x|}{2}\Big) \leq K^{-\frac{1}{p-1}}\Big(\frac{3R}{2}\Big) = o(R^{\frac{N}{p-1}}) \quad \text{for all } R < |z| < 2R$$

0 as $R \to \infty$. Thus, we may further estimate in (4.21) to deduce

$$\frac{1}{R^N} \int_{R\leq|y-x|\leq2R} |v(y)|dy = o(1) \to 0 \quad \text{as } R \to \infty. \tag{4.22}$$

If $p = 1$ we simply use the property (4.2) and (4.3) to estimate

$$\int_{R\leq|y-x|\leq2R} |v(y)|dy \leq \int_{R\leq|y-x|\leq2R} \Psi\Big(\frac{|y|}{2}\Big)\Psi^{-1}\Big(\frac{|y|}{2}\Big)|v(y)|dy$$

$$\leq \Psi^{-1}\Big(\frac{3R}{2}\Big) \int_{R\leq|y-x|\leq2R} \Psi\Big(\frac{|y|}{2}\Big)|v(y)|dy$$

$$\leq C\Psi^{-1}\Big(\frac{3R}{2}\Big) = o(R^N)$$

for all $R > |x|$ as $R \to \infty$. This shows that (4.22) also holds for $p = 1$. Hence, v satisfies the condition (4.6) in Theorem 4.1 with $\ell = 0$. The representation integral (4.19) follows now by Theorem 4.1. To prove the estimate (4.20), let $R > 0$ be such that $\int_{B_R} f(y)dy > 0$. Then, if $|x| \geq 1$ and $|y| < R$, we have $|x - y| \leq |x| + R \leq (R + 1)|x|$. Thus,

$$v(x) = c\int_{\mathbb{R}^N} \frac{f(y)}{|x-y|^{N-2m}}dy \geq c\int_{B_R} \frac{f(y)}{|x-y|^{N-2m}}dy$$

$$\geq c(R+1)^{2m-N}|x|^{2m-N} \int_{B_R} f(y)dy$$

$$\geq C|x|^{2m-N}.$$

□

4.3 An a Priori Estimate

We establish an estimate which holds for the general inequality (4.1).

Lemma 4.4 *Let $\psi \in C_c^\infty(\mathbb{R}^N)$ be such that supp $\psi \subset B_2$, $0 \leq \psi \leq 1$, $\psi \equiv 1$ on B_1. For $R > 2$ define $\varphi(x) = \psi^{2m}(x/R)$. Then, there exists $C > 0$ such that any nonnegative solution u of (4.1) satisfies*

$$\int_{\mathbb{R}^N} u\varphi dx \geq CR^{-N+2m}\left(\int_{\mathbb{R}^N} u^{\frac{p+q}{2}}\varphi dx\right)^2. \tag{4.23}$$

Proof It is easy to check that

$$\left|\Delta^m(\psi^{4m})\right| \leq C\psi^{2m} \quad \text{in } \mathbb{R}^N,$$

where $C > 0$ is a positive constant. This yields

$$\left|\Delta^m(\varphi^2)\right| \leq \frac{C}{R^{2m}}\varphi \quad \text{in } \mathbb{R}^N.$$

We multiply by φ^2 in (4.1) and integrate. Using the above estimate, we find

$$\int_{\mathbb{R}^N} \left(K(|x|) * u^p\right)u^q\varphi^2 \leq \pm\int_{\mathbb{R}^N} \varphi^2(\Delta^m u) = \pm\int_{\mathbb{R}^N} u\Delta^m(\varphi^2)$$

$$\leq \int_{\mathbb{R}^N} u\left|\Delta^m(\varphi^2)\right| \leq \frac{C}{R^{2m}}\int_{B_{2R}} u\varphi. \tag{4.24}$$

We next estimate the left-hand side of (4.24). By interchanging the variables and Hölder's inequality, one gets

$$\left(\int_{\mathbb{R}^N} \left(K(|x|) * u^p\right)u^q\varphi^2 dx\right)^2$$

$$= \left(\iint_{\mathbb{R}^N \times \mathbb{R}^N} K(|x-y|)u^p(x)u^q(y)\varphi^2(y)dxdy\right)^2$$

$$= \left(\iint_{\mathbb{R}^N \times \mathbb{R}^N} K(|x-y|)u^p(x)u^q(y)\varphi^2(y)dxdy\right) \times$$

$$\times \left(\iint_{\mathbb{R}^N \times \mathbb{R}^N} K(|x-y|)u^p(y)u^q(x)\varphi^2(x)dxdy\right)$$

$$\geq \left(\iint_{\mathbb{R}^N \times \mathbb{R}^N} K(|x-y|) u^{\frac{p+q}{2}}(x) u^{\frac{p+q}{2}}(y) \varphi(x) \varphi(y) dx dy \right)^2$$

$$\geq K(4R)^2 \left(\iint_{B_{2R} \times B_{2R}} u^{\frac{p+q}{2}}(x) u^{\frac{p+q}{2}}(y) \varphi(x) \varphi(y) dx dy \right)^2$$

$$\geq c R^{-2N} \left(\iint_{B_{2R} \times B_{2R}} u^{\frac{p+q}{2}}(x) u^{\frac{p+q}{2}}(y) \varphi(x) \varphi(y) dx dy \right)^2,$$

where $c > 0$ is a constant. Now, splitting the integrals according to x and y variables, we deduce

$$\int_{\mathbb{R}^N} \left(K(|x|) * u^p \right) u^q dx \geq c R^{-N} \left(\int_{\mathbb{R}^N} u^{\frac{p+q}{2}}(x) \varphi(x) dx \right)^2.$$

Using this last inequality in (4.24), we deduce (4.23).   □

4.4 The Inequality $-(-\Delta)^m u \geq (K * u^p) u^q$ in $\mathbb{R}^N$

In this section, we discuss the inequality

$$-(-\Delta)^m u \geq \left(K(|x|) * u^p \right) u^q \quad \text{in } \mathbb{R}^N. \tag{4.25}$$

Our main result in this case reads as follows.

Theorem 4.5 *Assume*

$$N, m \geq 1 \quad \text{and} \quad p + q \geq 2, \tag{4.26}$$

or

$$N > 2m \quad \text{and} \quad p \geq 1. \tag{4.27}$$

If $u \in C^{2m}(\mathbb{R}^N)$ is a nonnegative solution of (4.25), then $u \equiv 0$.

Before we prove Theorem 4.5, we establish the following general result.

Lemma 4.6 *Let $u \in C^{2m}(\mathbb{R}^N)$ be such that $u \geq 0$ and $(-\Delta)^m u \leq 0$ in $\mathbb{R}^N$. If*

$$\int_{B_R} u \, dx = o(R^N) \quad \text{as } R \to \infty, \tag{4.28}$$

then $u \equiv 0$.

Proof of Lemma 4.6 Let $u_j = (-\Delta)^j u$. We show by backward induction that $u_j \leq 0$ in $\mathbb{R}^N$. For $j = m$ this is true according to our hypothesis. Assume now that $u_m, u_{m-1}, \ldots, u_{j+1} \leq 0$ in $\mathbb{R}^N$ and prove that $u_j \leq 0$ in $\mathbb{R}^N$. Assume by contradiction that for some $x_0 \in \mathbb{R}^N$ one has $u_j(x_0) > 0$; without loosing the generality, we may assume $x_0 = 0$, so $u_j(0) > 0$.

Case 1: j is odd. From $u_{j+1} \leq 0$ in $\mathbb{R}^N$, we have $\Delta^{j+1}u \leq 0$ in $\mathbb{R}^N$. By taking the spherical average with respect to the origin, one deduces

$$\Delta(\overline{\Delta^j u}) = \Delta(\Delta^j \overline{u}) \leq 0 \quad \text{for all } r \geq 0.$$

By integration, one has $r^{N-1}\left(\Delta^j \overline{u}\right)' \leq 0$ for all $r \geq 0$. Hence,

$$\Delta^j \overline{u}(r) \leq \Delta^j \overline{u}(0) = -u_j(0) < 0 \quad \text{for all } r \geq 0.$$

By successive integration, one has

$$\Delta^{j-1}\overline{u}(r) \leq \Delta^{j-1}\overline{u}(0) + \frac{r^2}{2N}\Delta^j \overline{u}(0) \quad \text{for all } r \geq 0,$$

and then

$$\overline{u}(r) \leq \overline{u}(0) + \sum_{i=1}^{j} \frac{\Delta^i \overline{u}(0)r^{2i}}{\Pi_{k=1}^i[(2k)(N+2k-2)]} \quad \text{for all } r \geq 0.$$

The above inequality together with $\Delta^j \overline{u}(0) < 0$ yields $\overline{u}(r) \to -\infty$ as $r \to \infty$, which contradicts the fact that $u \geq 0$ in $\mathbb{R}^N$. Hence, $u_j \leq 0$ which finishes the proof in this case.

Case 2: j is even. Then $\Delta \overline{u}_j = \overline{\Delta u_j} = -\overline{u}_j \geq 0$. By integration one has $r^{N-1}\overline{u}_j'(r) \geq 0$ for all $r \geq 0$. Hence, $\overline{u}_j(r) \geq \overline{u}_j(0) > 0$ for all $r \geq 0$.

Let $\psi \in C_c^\infty(\mathbb{R}^N)$ be a cutoff function as in the statement of Lemma 4.4, namely, supp $\psi \subset B_2, 0 \leq \psi \leq 1, \psi \equiv 1$ on B_1. Then,

$$\int_{\mathbb{R}^N} u_j(x)\psi\left(\frac{x}{R}\right)dx \geq \int_{B_R} u_j(x)dx = C\int_0^R r^{N-1}\overline{u}_j(r)dr \geq CR^N \overline{u}_j(0).$$

On the other hand, by our hypothesis (4.28), we have

$$\int_{\mathbb{R}^N} u_j(x)\psi\left(\frac{x}{R}\right)dx = \int_{\mathbb{R}^N}(-\Delta)^j u(x)\psi\left(\frac{x}{R}\right)dx$$

$$= R^{-2j}\int_{\mathbb{R}^N} u(x)(-\Delta)^j\psi\left(\frac{x}{R}\right)dx$$

$$\leq CR^{-2j}\int_{B_{2R}} u(x)dx = o(R^{N-2j}) \quad \text{as } R \to \infty.$$

Combining the last two estimates, we find $CR^N \bar{u}_j(0) = o(R^{N-2j})$ as $R \to \infty$ which is a contradiction. Hence, $\bar{u}_j \leq 0$ in $\mathbb{R}^N$ which concludes our proof by induction.

$\square$

Proof of Theorem 4.5 We shall discuss separately the cases where (4.26) or (4.27) holds.

Case 1: $N, m \geq 1$ and $p + q \geq 2$. Let φ be as in Lemma 4.4. By Hölder's inequality, we have

$$\int_{\mathbb{R}^N} u\varphi \leq \left(\int_{\mathbb{R}^N} u^{\frac{p+q}{2}} \varphi \right)^{\frac{2}{p+q}} \left(\int_{\mathbb{R}^N} \varphi \right)^{1-\frac{2}{p+q}}$$

$$\leq CR^{N(1-\frac{2}{p+q})} \left(\int_{\mathbb{R}^N} u^{\frac{p+q}{2}} \varphi \right)^{\frac{2}{p+q}}.$$

Hence,

$$\left(\int_{\mathbb{R}^N} u^{\frac{p+q}{2}} \varphi \right)^2 \geq CR^{-N(p+q-2)} \left(\int_{\mathbb{R}^N} u\varphi \right)^{p+q}.$$

Using this last estimate in (4.23), we find

$$\int_{\mathbb{R}^N} u\varphi \leq CR^{N-\frac{2m}{p+q-1}} \quad \text{for all } R > 2.$$

In particular, since $\varphi = 1$ on B_R, we deduce

$$\int_{B_R} u \, dx = o(R^N) \quad \text{as } R \to \infty.$$

By Lemma 4.6 it now follows that $u \equiv 0$ which concludes our proof in this case.

Case 2: $N > 2m$ and $p \geq 1$. We apply Lemma 4.3 for $v = -u$. It follows in particular that $v = -u \geq 0$ which yields $u \equiv 0$.

$\square$

4.5   The Inequality $(-\Delta)^m u \geq (K * u^p)u^q$ in $\mathbb{R}^N$

In this section, we discuss the inequality

$$(-\Delta)^m u \geq \left(K(|x|) * u^p \right) u^q \quad \text{in } \mathbb{R}^N. \tag{4.29}$$

An important matter in the study of polyharmonic problems is whether nonnegative solutions of $(-\Delta)^m u \geq f(u)$ in $\mathbb{R}^N$ enjoy the so-called poly-superharmonic property, that is, whether $(-\Delta)^j u \geq 0$ in $\mathbb{R}^N$ for all $1 \leq j \leq m$. This has been shown to be true under some general conditions for nonlinearities $f(u)$.

In this section, we show that the poly-superharmonic property is still preserved in case of differential inequalities of type (4.29) that feature convolution terms. More generally, we show that the poly-superharmonic property still holds for polyharmonic systems of type

$$
\begin{cases}
(-\Delta)^m u \geq \big(K(|x|) * v^{p_1}\big)v^{q_1} \\[2mm]
(-\Delta)^m v \geq \big(L(|x|) * u^{p_2}\big)u^{q_2}
\end{cases}
\quad \text{in } \mathbb{R}^N,
\tag{4.30}
$$

Theorem 4.7 *Assume $N, m \geq 1$ and let Φ and Ψ satisfy (4.2). Suppose that (u, v) is a nonnegative solution of*
   where either

$$
p_1 + q_1 \geq 2 \quad \text{and} \quad p_2 + q_2 \geq 2,
\tag{4.31}
$$

or

$$
p_1, p_2 \geq 1.
\tag{4.32}
$$

Then, for all $1 \leq i \leq m$, we have

$$
(-\Delta)^i u \geq 0 \quad \text{and} \quad (-\Delta)^i v \geq 0 \quad \text{in } \mathbb{R}^N.
$$

Proof If $u \equiv 0$ (resp. $v \equiv 0$), then $v \equiv 0$ (resp. $u \equiv 0$). Hereafter, we assume that $u \not\equiv 0$ and $v \not\equiv 0$. The proof is divided into two steps.

Step 1: We have $(-\Delta)^{m-1}u \geq 0$ and $(-\Delta)^{m-1}v \geq 0$ in $\mathbb{R}^N$.
Assume by contradiction that there exists $x_0 \in \mathbb{R}^N$ such that $(-\Delta)^{m-1}u(x_0) < 0$ and without loosing any generality one may assume $x_0 = 0$. Let $u_i = (-\Delta)^i u$ and $v_i = (-\Delta)^i v$, $1 \leq i \leq m$ and denote by $\bar{u}(r)$, $\bar{v}(r)$ (resp $\bar{u}_i(r)$, $\bar{v}_i(r)$) the spherical average of u and v (resp u_i and v_i) on the sphere ∂B_r. Then

$$
\begin{cases}
-\Delta \bar{u} = \bar{u}_1, \qquad\quad -\Delta \bar{v} = \bar{v}_1, \\[1mm]
-\Delta \bar{u}_1 = \bar{u}_2, \qquad\quad -\Delta \bar{v}_1 = \bar{v}_2, \\[1mm]
\cdots\cdots\cdots\cdots \\[1mm]
-\Delta \bar{u}_{m-2} = \bar{u}_{m-1}, -\Delta \bar{v}_{m-2} = \bar{v}_{m-1}, \\[1mm]
-\Delta \bar{u}_{m-1} \geq \fint_{\partial B_r} \big(K(|x|) * v^{p_1}\big)u^{q_1} d\sigma \geq 0, \\[1mm]
-\Delta \bar{v}_{m-1} \geq \fint_{\partial B_r} \big(L(|x|) * u^{p_2}\big)v^{q_2} d\sigma \geq 0.
\end{cases}
\tag{4.33}
$$

From (4.33) one has $-\Delta \bar{u}_{m-1} \geq 0$ which yields $-r^{1-N}\left(r^{N-1}\bar{u}'_{m-1}\right)' \geq 0$ for all $r > 0$ and

$$\bar{u}_{m-1}(0) = (-\Delta)^{m-1}u(x_0) < 0.$$

By integration one gets

$$\bar{u}'_{m-1}(r) \leq 0 \quad \text{and} \quad \bar{u}_{m-1}(r) \leq \bar{u}_{m-1}(0) = u_{m-1}(x_0) < 0,$$

for all $r \geq 0$. We can rewrite the last estimate as

$$(-\Delta)^{m-1}\bar{u}(r) \leq (-\Delta)^{m-1}\bar{u}(0) < 0 \quad \text{for all } r \geq 0. \tag{4.34}$$

Case 1: m is odd. From (4.34) one has

$$\Delta^{m-1}\bar{u}(r) \leq \Delta^{m-1}\bar{u}(0) < 0 \quad \text{for all } r \geq 0.$$

Integrating twice the above inequality, we obtain

$$\Delta^{m-2}\bar{u}(r) \leq \Delta^{m-2}\bar{u}(0) + \frac{\Delta^{m-1}u(0)r^2}{2N} \quad \text{for all } r \geq 0$$

and proceeding further we deduce

$$\bar{u}(r) \leq \bar{u}(0) + \sum_{k=1}^{m-1} \frac{\Delta^k \bar{u}(0)}{\prod_{j=1}^{k}[(2j)(N+2j-2)]} r^{2k}. \tag{4.35}$$

Since $\Delta^{m-1}\bar{u}(0) = (-\Delta)^{m-1}u(x_0) < 0$, we deduce from (4.35) that

$$\bar{u}(r) \to -\infty \quad \text{as } r \to \infty,$$

which contradicts the fact that $u \geq 0$.

Case 2: m is even. Hence, $m \geq 2$. From (4.34) we find

$$\Delta^{m-1}\bar{u}(r) \geq \Delta^{m-1}\bar{u}(0) > 0 \quad \text{for all } r \geq 0.$$

In the same manner as we derived (4.35), it follows that for any $1 \leq i \leq m$, one has

$$\Delta^{m-i}\bar{u}(r) \geq \Delta^{m-i}\bar{u}(0) + \sum_{k=1}^{i-1} \frac{\Delta^{m-i+k}\bar{u}(0)}{\prod_{j=1}^{k}[(2j)(N+2j-2)]} r^{2k}.$$

In particular, for $i = m$ we find

$$\bar{u}(r) \ge \bar{u}(0) + \sum_{k=1}^{m-1} \frac{\Delta^k \bar{u}(0)}{\Pi_{j=1}^k [(2j)(N+2j-2)]} r^{2k}.$$

Since $\Delta^{m-1}\bar{u}(0) > 0$ and $m \ge 2$, it follows from the above estimate that

$$\bar{u}(r) \ge C_1 r^{2(m-1)} - C_2 \quad \text{for all } r \ge 0, \tag{4.36}$$

for some constants $C_1, C_2 > 0$.

Case 2a: Assume that (4.31) holds. Let φ be as in Lemma 4.4. In the same way as in the proof of (4.23), we have

$$\begin{cases} \displaystyle\int_{\mathbb{R}^N} u\varphi \, dx \ge C R^{-N+2m} \left(\int_{\mathbb{R}^N} v^{\frac{\tau}{2}} \varphi \, dx \right)^2, \\[4mm] \displaystyle\int_{\mathbb{R}^N} v\varphi \, dx \ge C R^{-N+2m} \left(\int_{\mathbb{R}^N} u^{\frac{\theta}{2}} \varphi \, dx \right)^2, \end{cases} \tag{4.37}$$

where

$$\tau = p_1 + q_1 \ge 2 \quad \text{and} \quad \theta = p_2 + q_2 \ge 2.$$

By Hölder's inequality, we find

$$\begin{cases} \displaystyle\left(\int_{\mathbb{R}^N} v^{\frac{\tau}{2}} \varphi \, dx \right)^2 \ge C R^{-N(\tau-2)} \left(\int_{\mathbb{R}^N} v\varphi \, dx \right)^{\tau}, \\[4mm] \displaystyle\left(\int_{\mathbb{R}^N} u^{\frac{\theta}{2}} \varphi \, dx \right)^2 \ge C R^{-N(\theta-2)} \left(\int_{\mathbb{R}^N} u\varphi \, dx \right)^{\theta}. \end{cases} \tag{4.38}$$

From (4.37) and (4.38), we deduce

$$\begin{cases} \displaystyle\int_{\mathbb{R}^N} u\varphi \, dx \ge C R^{-N+2m-N(\tau-2)} \left(\int_{\mathbb{R}^N} v\varphi \, dx \right)^{\tau}, \\[4mm] \displaystyle\int_{\mathbb{R}^N} v\varphi \, dx \ge C R^{-N+2m-N(\theta-2)} \left(\int_{\mathbb{R}^N} u\varphi \, dx \right)^{\theta}. \end{cases} \tag{4.39}$$

We use the second estimate of (4.39) in the first one to obtain

$$\int_{\mathbb{R}^N} u\psi \, dx \ge C R^{-N+2m-N(\tau-2)-(N-2m)\tau-N(\theta-2)\tau} \left(\int_{\mathbb{R}^N} u\varphi \, dx \right)^{\tau\theta},$$

which we arrange as

$$CR^{N-\frac{2m(1+\tau)}{\tau\theta-1}} \geq \int_{B_R} u dx.$$

Using (4.36) we find

$$CR^{N-\frac{2m(1+\tau)}{\tau\theta-1}} \geq \int_{B_R} u dx = \sigma_N \int_0^R r^{N-1}\bar{u} dr$$

$$\geq C_3 R^{N+2(m-1)} - C_4 R^N \quad \text{for } R > 1 \text{ large,}$$

where $C_3, C_4 > 0$ and σ_N denotes the surface area of the unit sphere in $\mathbb{R}^N$. Comparing the exponents of R in the above inequality, we raise a contradiction, since $C_3 > 0$. This finishes the proof of Step 1 in this case.

Case 2b: Assume that (4.32) holds. From (4.36) one can find $r_0 > 0$ and a constant $c > 0$ such that

$$\bar{u}(r) \geq cr^{2(m-1)} \quad \text{for all } r \geq r_0. \tag{4.40}$$

Using the fact that $r^N K(r) \to \infty$ as $r \to \infty$, by taking $r_0 > 1$ large enough, we may also assume that

$$r^N K\left(\frac{r}{2}\right) \geq 1 \quad \text{for all } r \geq r_0. \tag{4.41}$$

To raise a contradiction, we next return to condition (4.3) for u (in which we replace p with p_2). From (4.40)–(4.41), co-area formula and Jensen's inequality, we obtain

$$\infty > \int_{|y|>r_0} K\left(\frac{|y|}{2}\right) u^{p_2}(y) dy = \int_{r_0}^\infty \int_{|y|=r} K\left(\frac{|y|}{2}\right) u^{p_2}(y) d\sigma(y)\, dr$$

$$= \int_{r_0}^\infty K\left(\frac{r}{2}\right) \int_{|y|=r} u^{p_2}(y) d\sigma(y)\, dr$$

$$\geq \sigma_N \int_{r_0}^\infty r^{N-1} K\left(\frac{|r|}{2}\right) \bar{u}^{p_2}(r) dr$$

$$\geq C \int_{r_0}^\infty r^{2p_2(m-1)-1} r^N K\left(\frac{|r|}{2}\right) dr$$

$$\geq C \int_{r_0}^\infty r^{2p_2(m-1)-1} dr = \infty,$$

which is a contradiction and concludes the proof in Step 1.

Step 2: We have $(-\Delta)^{m-i}u \geq 0$ and $(-\Delta)^{m-i}v \geq 0$ in $\mathbb{R}^N$ for any $1 \leq i \leq m$.

From Step 1 we know that $(-\Delta)^{m-1}u \geq 0$, $(-\Delta)^{m-1}v \geq 0$ in $\mathbb{R}^N$. Letting $u_{m-2} = (-\Delta)^{m-2}u$ and $v_{m-2} = (-\Delta)^{m-2}v$, we want to show next that $u_{m-2} \geq 0$ and $v_{m-2} \geq 0$ in $\mathbb{R}^N$. Suppose to the contrary that there exists $x_0 \in \mathbb{R}^N$ so that $u_{m-2}(x_0) < 0$. We next take the spherical average with respect to spheres centred at x_0 and proceed as in Step 1 by discussing separately the cases m is odd and m is even in order to raise a contradiction. Thus, $(-\Delta)^{m-2}u \geq 0$, $(-\Delta)^{m-2}v \geq 0$ in $\mathbb{R}^N$. We proceed further until we get $-\Delta u \geq 0$, $-\Delta v \geq 0$ in $\mathbb{R}^N$.

$\square$

Corollary 4.8 *Assume* $N, m \geq 1$ *and either* $p + q \geq 2$ *or* $p \geq 1$.
If $u \in C^{2m}(\mathbb{R}^N)$ *is a nonnegative solution of* (4.29), *then, for all* $1 \leq j \leq m$, *we have*

$$(-\Delta)^j u \geq 0 \quad in \ \mathbb{R}^N.$$

Proof Let $K = L$ and $(p_1, q_1) = (p_2, q_2) = (p, q)$. Suppose u is a nonnegative solution of (4.29). If $u \equiv 0$, then the conclusion clearly holds. If $u \not\equiv 0$, then (u, u) is a nonnegative solution of (4.30). By Theorem 4.7 we see that, for all $1 \leq i \leq m$, $(-\Delta)^i u \geq 0$ in $\mathbb{R}^N$. $\square$

Our next goal is to establish the following nonexistence result.

Theorem 4.9 *Assume* $N, m \geq 1$ *and let* $u \in C^{2m}(\mathbb{R}^N)$ *be a nonnegative solution of* (4.29).

(i) *If* $1 \leq N \leq 2$ *and* $(p \geq 1$ *or* $p + q \geq 2)$, *then* $u \equiv 0$.
(ii) *If* $N > 2m$ *and one of the following conditions holds:*

   (ii1) $(p \geq 1$ *or* $p + q \geq 2)$ *and* $\int_{|y| > 1} |y|^{-p(N-2m)} K(|y|)dy = \infty$;
   (ii2) $p + q \geq 2$ *and* $\limsup\limits_{r \to \infty} r^{2N - (N-2m)(p+q)} K(r) > 0$;

   then, $u \equiv 0$.

Proof

(i) By Corollary 4.8 we see that, for $1 \leq j \leq m$, $(-\Delta)^j u \geq 0$ in $\mathbb{R}^N$. In particular $-\Delta u \geq 0$ in $\mathbb{R}^N$. Since $N = 1, 2$, it is well known that a nonnegative superharmonic function is constant. Thus, $u = c$ in $\mathbb{R}^N$. By (4.29) it follows that $(K(|x|) * u^p)u^q = 0$ in $\mathbb{R}^N$. This clearly yields $u = 0$ in $\mathbb{R}^N$; otherwise, there would exist $x_0 \in \mathbb{R}^N$ such that $u(x_0) > 0$ and hence $(K(|x|) * u^p)u^q > 0$ at x_0.

(ii) By estimate (4.20) in Proposition 4.3, we find $u \geq c|x|^{2m-N}$ in $\mathbb{R}^n \setminus B_1$, for some $c > 0$. Thus, by the hypothesis (ii1) in Theorem 4.9, we find

$$\int_{|y|>1} K\left(\frac{|y|}{2}\right) u^p(y) dy \geq c \int_{|y|>1} |y|^{-p(N-2m)} K(|y|) dy = \infty,$$

which contradicts (4.3).

The proof of part (ii2) follows from the proof of Theorem 4.11.   □

In the case $K(r) = r^{-\alpha}$, $\alpha \in (0, N)$, we obtain the following theorem which retrieves the result in Corollary 2.16 related to the inequality (2.55) for $m = 2$.

Theorem 4.10 *Assume $N > 2m$, $m \geq 1$, $\alpha \in (0, N)$, $p \geq 1$ and $q > 1$. Then, the inequality*

$$(-\Delta)^m u \geq \left(|x|^{-\alpha} * u^p\right) u^q \quad in \ \mathbb{R}^N \tag{4.42}$$

has nonnegative nontrivial solutions if and only if

$$\min\{p, q\} > \frac{N - \alpha}{N - 2m} \quad and \quad p + q > \frac{2N - \alpha}{N - 2m}. \tag{4.43}$$

Proof Assume first that (4.42) has a nonnegative solution $u \not\equiv 0$. Then, by estimate (4.20) in Corollary 4.3, we have $u \geq c|x|^{2m-N}$ in $\mathbb{R}^N \setminus B_1$, where $c > 0$ is a constant. It is easy to check that the condition $p > (N - \alpha)/(N - 2m)$ and (4.43)$_2$ follow from Theorem 4.9 with $\Psi(r) = r^{-\alpha}$.

It remains to prove that $q > (N - \alpha)/(N - 2m)$. If $\alpha \geq 2m$ then this is clearly true, since $q > 1$. Assume next that $\alpha < 2m$.

For $x \in \mathbb{R}^N \setminus B_1$ and $1 < |y| < 2$, we have $|x - y| \leq 3|x|$. Thus,

$$|x|^{-\alpha} * u^p \geq \int_{\mathbb{R}^N} \frac{f(y)}{|x - y|^\alpha} dy \geq \int_{1<|y|<2} \frac{u^p(y)}{|x - y|^\alpha} dy$$

$$\geq \int_{1<|y|<2} \frac{u^p(y) dy}{(3|x|)^\alpha} \geq C|x|^{-\alpha}.$$

Thus, $|x|^{-\alpha} * u^p \geq C|x|^{-\alpha}$ for all $x \in \mathbb{R}^N \setminus B_1$ and u satisfies

$$(-\Delta)^m u \geq c|x|^{-\alpha} u^q \quad in \ \mathbb{R}^N \setminus B_1,$$

for some $c > 0$. Let $\psi \in C_c^\infty(\mathbb{R}^N)$ be a cutoff function such that supp $\psi \subset B_2$, $0 \leq \psi \leq 1$, $\psi \equiv 1$ on B_1. Then, letting $\phi_R(x) = \psi(x/R)$, by Hölder's inequality, one has

$$\int_{\mathbb{R}^N \setminus B_1} |x|^{-\alpha} u^p \phi_R dx$$

$$\leq \int_{\mathbb{R}^N \setminus B_1} (-\Delta)^m u(x)\phi_R dx = \int_{\text{supp}(\nabla\phi_R)} u(x)(-\Delta)^m \phi_R dx$$

$$\leq \left(\int_{\text{supp}(\nabla\phi_R)} |x|^{-\alpha} u^p \phi_R \right)^{1/q} \left(\int_{\mathbb{R}^N \setminus B_1} |x|^{\alpha/(q-1)} \frac{|(-\Delta)^m \phi_R|^{q'}}{\phi_R^{1/(q-1)}} \right)^{1/q'},$$

$$(4.44)$$

which yields

$$\int_{\mathbb{R}^N \setminus B_1} |x|^{-\alpha} u^p \phi_R \leq \int_{\mathbb{R}^N \setminus B_1} |x|^{\alpha/(q-1)} \frac{|(-\Delta)^m \phi_R|^{q'}}{\phi_R^{1/(q-1)}} \leq C R^{N+(\alpha-2mq)(q-1)},$$

$$(4.45)$$

where $C > 0$ is a constant independent of $R > 1$.

If $N + (\alpha - 2mq)(q - 1) < 0$, that is, if $q < (N - \alpha)/(n - 2m)$ then, letting $R \to \infty$ in (4.45), we deduce $u \equiv 0$ on $\mathbb{R}^N \setminus B_1$, contradiction.

If $N + (\alpha - 2mq)(q - 1) = 0$, that is, if $q = (N - \alpha)/(n - 2m)$ then, from (4.45) we deduce $|x|^{-\alpha} u^q \in L^1(\mathbb{R}^N \setminus B_1)$. This shows that

$$\int_{\text{supp}(\nabla\phi_R)} |x|^{-\alpha} u^p \phi_R \leq \int_{B_{2R} \setminus B_R} |x|^{-\alpha} u^p \to 0 \quad \text{as } R \to \infty,$$

and from (4.44) we find again $u \equiv 0$ on $\mathbb{R}^N \setminus B_1$, contradiction. Thus, if $\alpha < 2m$ and $1 < q \leq (N - \alpha)/(N - 2m)$, then there are no nonnegative solutions apart from the trivial one. Hence, $q > (N - \alpha)/(N - 2m)$.

Assume now that (4.43) holds and let us construct a positive solution to (4.42). First, we write (4.43) in the form

$$(N - 2m)(p + q - 1) > N - \alpha + 2m \quad (N - 2m)p > N - \alpha \quad \text{and} \quad (N - 2m)(q - 1) > 2m - \alpha.$$

Thus, we can choose $\kappa \in (0, N - 2m)$ such that

$$\begin{cases} \kappa(p + q - 1) > N - \alpha + 2m, \\ \kappa p > N - \alpha, \\ \kappa(q - 1) > 2m - \alpha, \\ p\kappa \neq N. \end{cases} \quad (4.46)$$

For $a \geq 0$ we define

$$F(a, x) = (-\Delta)^m \left\{ (a + |x|^2)^{-\kappa/2} \right\} \quad \text{for all } x \in \mathbb{R}^N \setminus \{0\}.$$

Then,

$$F(a, x) = (a + |x|^2)^{-\frac{\kappa}{2}-2m} \sum_{j=0}^{m} b_j(a)|x|^{2j} \quad \text{for all } x \in \mathbb{R}^N,$$

where $b_j(a) \in \mathbb{R}$. In particular, for $a = 0$ we find

$$F(0, x) = |x|^{-\kappa-4m} \sum_{j=0}^{m} b_j(0)|x|^{2j} \quad \text{for all } x \in \mathbb{R}^N \setminus \{0\}. \tag{4.47}$$

On the other hand, by direct computation, one has

$$F(0, x) = (-\Delta)^m \left\{ |x|^{-\kappa} \right\} = \prod_{j=1}^{m} \left[(\kappa + 2j - 2)(N - \kappa - 2j) \right] |x|^{-\kappa-2m} > 0,$$

$$\tag{4.48}$$

since $0 < \kappa < N - 2m$. Comparing (4.47) and (4.48), we find $b_{2m}(0) > 0$. By the continuous dependence on the data, we can find now $a > 0$ such that $b_{2m}(a) > 0$. Also,

$$\lim_{|x| \to \infty} \frac{F(a, x)}{|x|^{-k-2m}} = b_{2m}(a) > 0.$$

Thus, there exist $c > 0$ and $R > 1$ such that $F(a, x) \geq c|x|^{-\kappa-2m}$ for $x \in \mathbb{R}^N \setminus B_R$.

Let now $v(x) = (a + |x|^2)^{-\kappa/2}$, where $a > 0$ satisfies $b_{2m}(a) > 0$. By the above estimates, we have

$$(-\Delta)^m v \geq c|x|^{-\kappa-2m} \quad \text{in } \mathbb{R}^N \setminus B_R. \tag{4.49}$$

Let $\varphi \in C_c^1(\mathbb{R}^N)$, $0 \leq \varphi \leq 1$ such that supp $\varphi \subset B_{2R}$ and $\varphi \equiv 1$ on B_R. For $M > 1$ define

$$V(x) = v(x) + M\gamma_0 \int_{\mathbb{R}^N} \frac{\varphi(y)}{|x - y|^{N-2m}} dy \quad \text{for all } x \in \mathbb{R}^N,$$

where $\gamma_0 > 0$ is a normalising constant such that

$$(-\Delta)^m \left(\gamma_0 |x|^{2m-N} \right) = \delta_0 \quad \text{in } \mathscr{D}'(\mathbb{R}^N),$$

and δ_0 denotes the Dirac mass concentrated at the origin.

Thus, $V \in C^{2m}(\mathbb{R}^N)$, $V > 0$ in $\mathbb{R}^N$ and from (4.49) we have

$$(-\Delta)^m V \geq c|x|^{-\kappa-2m} \quad \text{in } \mathbb{R}^N \setminus B_R. \tag{4.50}$$

Also, by taking $M > 1$ large enough, we have

$$(-\Delta)^m V = (-\Delta)^m v + M > 0 \quad \text{in } \overline{B}_R. \tag{4.51}$$

Observe that for $x \in \mathbb{R}^N \setminus B_{4R}$ and $y \in B_{2R}$, we have $|x - y| \geq |x| - |y| \geq |x|/2$. Thus,

$$\int_{\mathbb{R}^N} \frac{\varphi(y)}{|x - y|^{N-2m}} dy = \int_{B_{2R}} \frac{\varphi(y)}{|x - y|^{N-2m}} dy$$

$$\leq 2^{N-2m}|x|^{2m-N} \int_{B_{2R}} \varphi(y) dy$$

$$\leq C|x|^{2m-N}.$$

Using this estimate in the definition of V together with $0 < \kappa < N - 2m$, it follows that

$$V(x) \leq C_0 |x|^{-\kappa} \quad \text{for all } x \in \mathbb{R}^N \setminus B_{R/2}, \tag{4.52}$$

for some constant $C_0 > 0$.

We next evaluate the convolution term $(|x|^{-\alpha} * V^p)V^q$ and indicate how to construct a positive solution to (4.42). Using (4.52) we can apply Lemma 2.12 for $f = V^p$, $\beta = \kappa p > N - \alpha$ and $\rho = R/2$. It follows that for any $x \in \mathbb{R}^N \setminus B_R$, we have

$$(|x|^{-\alpha} * V^p)V^q \leq c|x|^{-\kappa q} \int_{\mathbb{R}^N} \frac{V(y)^p dy}{|x - y|^\alpha} \leq C \begin{cases} |x|^{N-\alpha-\kappa(p+q)} & \text{if } \kappa p < N, \\ |x|^{-\alpha-\kappa q} & \text{if } \kappa p > N. \end{cases}$$

Using this last estimate together with (4.50) and (4.46)$_1$, (4.46)$_3$ we deduce

$$(-\Delta)^m V \geq C_1 (|x|^{-\alpha} * V^p)V^q \quad \text{in } \mathbb{R}^N \setminus B_R, \tag{4.53}$$

for some $C_1 > 0$. Since $(-\Delta)^m V$ and $(|x|^{-\alpha} * V^p)V^q$ are continuous and positive functions on the compact $\overline{B}_R$ (see (4.51)), one can find $C_2 > 0$ such that

$$(-\Delta)^m V \geq C_2 (|x|^{-\alpha} * V^p)V^q \quad \text{in } B_R. \tag{4.54}$$

Thus, letting $C = \min\{C_1, C_2\} > 0$ and $U = C^{1/(p+q-1)}V$, it follows that $U \in C^{2m}(\mathbb{R}^N)$ is positive and that from (4.53)–(4.54), one has

$$(-\Delta)^m U \geq (|x|^{-\alpha} * U^p)U^q \quad \text{in } \mathbb{R}^N,$$

which concludes our proof. $\qquad\qquad\qquad\square$

4.6   Polyharmonic Systems with Convolution Terms

In this section, we discuss the corresponding systems associated with (4.29), namely,

$$(-\Delta)^m u_i \geq \sum_{j=1}^n e_{ij} \left(K_{ij}(|x|) * u_j^{p_{ij}} \right) u_j^{q_{ij}} \quad \text{in } \mathbb{R}^N, \ 1 \leq i \leq n \qquad (4.55)$$

and

$$(-\Delta)^m u_i \geq \sum_{j=1}^n e_{ij} \left(K_{ij}(|x|) * u_j^{p_{ij}} \right) u_i^{q_{ij}} \quad \text{in } \mathbb{R}^N, \ 1 \leq i \leq n \qquad (4.56)$$

where $N, m \geq 1$, $p_{ij} \geq 1$, $q_{ij} > 0$ and (e_{ij}) is the adjacency matrix, i.e. e_{ij} satisfies

$$e_{ij} = 0 \text{ or } 1 \quad \text{and} \quad e_{ij} = e_{ji} \quad \text{for } i, j \in \{1, \ldots, n\}.$$

By a nonnegative solution of (4.55) (resp. (4.56)), we understand a n-component function $u = (u_1, u_2, \ldots, u_n)$ with $u_j \in C^{2m}(\mathbb{R}^N)$, $u_j \geq 0$, such that

$$\sum_{i=1}^n \int_{|y|>1} K_{ij}\left(\frac{|y|}{2}\right) u_j^{p_{ij}}(y) dy < \infty \qquad (4.57)$$

and u satisfies (4.55) (resp. (4.56)) pointwise.

The main result for (4.55) is the following:

Theorem 4.11 *Assume $N > 2m$, $m \geq 1$ and let $(u_1, \ldots, u_n)$ be a nonnegative solution of (4.55). If there exist $k, \ell \in \{1, \ldots, n\}$ (not necessarily distinct) such that*

$$e_{k\ell} = e_{\ell k} = 1,$$

$$p_{k\ell} + q_{k\ell} \geq 2, \quad p_{\ell k} + q_{\ell k} \geq 2, \qquad (4.58)$$

$$\limsup_{r \to \infty} \min \left\{ r^{2N-(N-2m)(p_{k\ell}+q_{k\ell})} K_{k\ell}(r), \ r^{2N-(N-2m)(p_{\ell k}+q_{\ell k})} K_{\ell k}(r) \right\} > 0, \qquad (4.59)$$

then $u_k \equiv u_\ell \equiv 0$.

Proof Let $a > 0$ denote the positive limit in (4.59). Thus, one can find an increasing sequence $\{R_i\} \subset (0, \infty)$ that tends to infinity and such that for all $i \geq 1$ one has

$$\min \left\{ R_i^{2N-(N-2m)(p_{k\ell}+q_{k\ell})} K_{k\ell}(R_i), \ r^{2N-(N-2m)(p_{\ell k}+q_{\ell k})} K_{\ell k}(R_i) \right\} > \frac{a}{2}. \qquad (4.60)$$

By Theorem 4.7 we deduce that u_k and u_ℓ are poly-superharmonic. Further, by estimate (4.20) in Proposition 4.3, there exists $c > 0$ such that

$$u_k(x), u_\ell(x) \geq c|x|^{2m-N} \quad \text{in } \mathbb{R}^N \setminus B_1. \tag{4.61}$$

Let ϕ be the positive eigenfunction of $-\Delta$ in the unit ball $\overline{B}_1$ corresponding to the eigenvalue $\lambda_1 > 0$. We normalise ϕ such that $0 \leq \phi \leq 1$ in B_1 and $\max_{\overline{B}_1} \phi(x) = 1$. Let $\varphi_i(x) = \phi(x/R_i)$. Multiplying by φ_i in the inequality of (4.55) that corresponds to u_k, we find

$$
\begin{aligned}
\int_{B_{R_i}} \left(K_{k\ell}(|x|) * u_\ell^{p_{k\ell}}\right) u_\ell^{q_{k\ell}} \varphi_i &\leq \int_{B_{R_i}} (-\Delta)^m u_k \varphi_i \\
&= \int_{B_{R_i}} (-\Delta)^{m-1} u_k (-\Delta)\varphi_i + \int_{\partial B_{R_i}} (-\Delta)^{m-1} u_k \frac{\partial \varphi_i}{\partial n} \\
&\leq \frac{\lambda_1}{R_i^2} \int_{B_{R_i}} (-\Delta)^{m-1} u_k \varphi_i,
\end{aligned}
$$

where we used $(-\Delta)^{m-1} u_k \geq 0$ by Theorem 4.7 and that, by Hopf lemma, $\partial \varphi_i / \partial n < 0$ on ∂B_{R_i}. Proceeding further one finds

$$\int_{B_{R_i}} \left(K_{k\ell}(|x|) * u_\ell^{p_{k\ell}}\right) u_\ell^{q_{k\ell}} \varphi_i \leq \left(\frac{\lambda_1}{R_i^2}\right)^m \int_{B_{R_i}} u_k \varphi_i. \tag{4.62}$$

Let us next estimate the integral in the left-hand side of (4.62). If $x \in B_{R_i}$, then one has

$$K_{k\ell}(|x|) * u_\ell^{p_{k\ell}} \geq \int_{B_{R_i}} K_{k\ell}(|x-y|) u_\ell^{p_{k\ell}}(y) dy \geq K_{k\ell}(2R_i) \int_{B_{R_i}} u_\ell^{p_{k\ell}}(y) dy.$$

Thus, by the fact that $0 \leq \varphi_i \leq 1$, one has

$$\int_{B_{R_i}} \left(K_{k\ell}(|x|) * u_\ell^{p_{k\ell}}\right) u_\ell^{q_{k\ell}} \varphi_i \geq K_{k\ell}(2R_i) \left(\int_{B_{R_i}} u_\ell^{p_{k\ell}} \varphi_i\right) \left(\int_{B_{R_i}} u_\ell^{q_{k\ell}} \varphi_i\right). \tag{4.63}$$

Combining (4.63) with (4.62), we obtain

$$K_{k\ell}(2R_i) \left(\int_{B_{R_i}} u_\ell^{p_{k\ell}} \varphi_i\right) \left(\int_{B_{R_i}} u_\ell^{q_{k\ell}} \varphi_i\right) \leq \left(\frac{\lambda_1}{R_i^2}\right)^m \int_{B_{R_i}} u_k \varphi_i. \tag{4.64}$$

Let $\tau = p_{k\ell} + q_{k\ell} \geq 2$. By Hölder's inequality, we have

$$\left(\int_{B_{R_i}} u_\ell^{\frac{\tau}{2}} \varphi_i\right)^2 \le \left(\int_{B_{R_i}} u_\ell^{p_{k\ell}} \varphi_i\right)\left(\int_{B_{R_i}} u_\ell^{q_{k\ell}} \varphi_i\right). \tag{4.65}$$

Again by Hölder's inequality, we derive

$$\int_{B_{R_i}} u_\ell \varphi_i \le \left(\int_{B_{R_i}} u_\ell^{\frac{\tau}{2}} \varphi_i\right)^{\frac{2}{\tau}}\left(\int_{B_{R_i}} \varphi_i\right)^{1-\frac{2}{\tau}} \le C R_i^{N\left(1-\frac{2}{\tau}\right)}\left(\int_{B_{R_i}} u_\ell^{\frac{\tau}{2}} \varphi_i\right)^{\frac{2}{\tau}},$$

and hence

$$\left(\int_{B_{R_i}} u_\ell \varphi_i\right)^\tau \le C R_i^{N(\tau-2)}\left(\int_{B_{R_i}} u_\ell^{\frac{\tau}{2}} \varphi_i\right)^2. \tag{4.66}$$

By (4.64), (4.65) and (4.66), we have

$$R_i^{2m-N(\tau-2)} K_{k\ell}(2R_i)\left(\int_{B_{R_i}} u_\ell \varphi_i\right)^\tau \le C \int_{B_{R_i}} u_k \varphi_i,$$

which we write it as

$$R_i^{2N-(N-2m)\tau} K_{k\ell}(2R_i)\left(\int_{B_{R_i}} u_\ell \varphi_i\right)^\tau \le C R_i^{2m(\tau-1)} \int_{B_{R_i}} u_k \varphi_i$$

Using (4.60) it follows that for $i \ge 1$ large enough, we have

$$\frac{a}{2}\left(\int_{B_{R_i}} u_\ell \varphi_i\right)^{p_{k\ell}+q_{k\ell}} \le C R_i^{2m(p_{k\ell}+q_{k\ell}-1)} \int_{B_{R_i}} u_k \varphi_i. \tag{4.67}$$

Similarly, we have

$$\frac{a}{2}\left(\int_{B_{R_i}} u_k \varphi_i\right)^{p_{\ell k}+q_{\ell k}} \le C R_i^{2m(p_{\ell k}+q_{\ell k}-1)} \int_{B_{R_i}} u_\ell \varphi_i. \tag{4.68}$$

Multiplying (4.67) with (4.68) and using the fact that

$$\left(\int_{B_{R_i}} u_k \varphi_i\right)\left(\int_{B_{R_i}} u_\ell \varphi_i\right) > 0 \text{ for large } i,$$

we have

$$\left(\int_{B_{R_i}} u_\ell \varphi_i\right)^{p_{k\ell}+q_{k\ell}-1} \left(\int_{B_{R_i}} u_k \varphi_i\right)^{p_{\ell k}+q_{\ell k}-1} \leq C R_i^{2m(p_{k\ell}+q_{k\ell}-1)} R_i^{2m(p_{\ell k}+q_{\ell k}-1)}.$$

From here we deduce that there exists a subsequence $\{R_i\}$ (still denoted in the following by $\{R_i\}$) such that[1]

- either $\left(\int_{B_{R_i}} u_\ell \varphi_i\right)^{p_{k\ell}+q_{k\ell}-1} \leq C R_i^{2m(p_{k\ell}+q_{k\ell}-1)}$;

- or $\left(\int_{B_{R_i}} u_k \varphi_i\right)^{p_{\ell k}+q_{\ell k}-1} \leq C R_i^{2m(p_{\ell k}+q_{\ell k}-1)}$.

Assume the second assertion holds. This yields

$$\int_{B_{R_i}} u_k \varphi_i \leq C R_i^{2m}.$$

Using this last estimate in (4.62), we deduce $\left(K_{k\ell}(|x|) * u_\ell^{p_{k\ell}}\right) u_\ell^{q_{k\ell}} \in L^1(\mathbb{R}^N)$, and so

$$\int_{B_{R_i} \setminus B_{R_i/2}} \left(K_{k\ell}(|x|) * u_\ell^{p_{k\ell}}\right) u_\ell^{q_{k\ell}} \varphi_i \to 0 \quad \text{as } i \to \infty. \tag{4.69}$$

We may estimate the above integral as we did in (4.63) to obtain

$$\int_{B_{R_i} \setminus B_{R_i/2}} \left(K_{k\ell}(|x|) * u_\ell^{p_{k\ell}}\right) u_\ell^{q_{k\ell}} \varphi_i \geq K_{k\ell}(2R_i) \left(\int_{B_{R_i} \setminus B_{R_i/2}} u_\ell^{p_{k\ell}} \varphi_i\right) \left(\int_{B_{R_i} \setminus B_{R_i/2}} u_\ell^{q_{k\ell}} \varphi_i\right).$$

Finally, we use (4.61) in the above inequality to deduce

$$\int_{B_{R_i} \setminus B_{R_i/2}} \left(K_{k\ell}(|x|) * u_\ell^{p_{k\ell}}\right) u_\ell^{q_{k\ell}} \geq C R_i^{2N-(N-2m)(p_{k\ell}+q_{k\ell})} \Psi_{k\ell}(2R_i) > CL > 0,$$

for large i, which contradicts (4.69) and concludes our proof.   □

Theorem 4.11 states that if k is adjacent to ℓ and (4.58)–(4.59) hold, then $u_k \equiv u_\ell \equiv 0$. We immediately obtain the following Liouville-type result:

Corollary 4.12 *Make the same assumptions as in Theorem 4.11. In particular, suppose that (4.58) and (4.59) hold for every pair (k, ℓ) such that $e_{k\ell} = e_{\ell k} = 1$. If each connected component of the network has more than two nodes, then the only*

---

[1] We make use of the following basic argument: if $a_i b_i \leq x_i y_i$, then either $a_i \leq x_i$ or $b_i \leq y_i$ (we argue by contradiction to prove this fact). Since $i \geq 1$ can be any (large) positive integer, along a subsequence, we have either $a_i \leq x_i$ or $b_i \leq y_i$.

nonnegative solution of (4.55) is $(0, \ldots, 0)$. *In particular, if the network has only one connected component and* $n \geq 2$, *then the only nonnegative solution of (4.55) is* $(0, \ldots, 0)$.

The main result regarding the system (4.56) is the following.

Theorem 4.13 *Assume* $N > 2m$, $m \geq 1$ *and let* $(u_1, \ldots, u_n)$ *be a nonnegative solution of (4.56). If there exist* $k, \ell \in \{1, \ldots, n\}$ *(not necessarily distinct) such that*

$$e_{k\ell} = e_{\ell k} = 1,$$

$$p_{k\ell}, q_{k\ell} \geq 1, \quad p_{\ell k}, q_{\ell k} \geq 1, \tag{4.70}$$

and (4.59) holds, then $u_k \equiv 0$ *or* $u_\ell \equiv 0$ *(or both).*

Proof The proof of Theorem 4.13 can be carried out in the same way as above. The only difference is that we cannot apply Theorem 4.7 to derive that u_k and u_ℓ satisfy (4.61). Instead, we apply Corollary 4.3 to deduce that u_k and u_ℓ are poly-superharmonic. Further, by the estimate, (4.20) one has that (4.61) holds. From now on, we follow line by line the above proof. □

Assume next that the adjacency matrix (e_{ij}) is given by

$$e_{ij} = \begin{cases} 1 & \text{if } i \neq j, \\ 0 & \text{if } i = j. \end{cases} \tag{4.71}$$

From Theorem 4.11 and Theorem 4.13, we find:

Corollary 4.14 *Suppose* $N > 2m$, $m \geq 1$ *and that* (e_{ij}) *is defined by (4.71).*

(i) *Assume (4.58)-(4.59). Then the only nonnegative solution of (4.55) is*

$$(u_1, \ldots, u_n) = (0, \ldots, 0).$$

(ii) *Assume (4.59)-(4.70). Then all nonnegative solutions of (4.56) are of the form*

$$(u_1, \ldots, u_n) = (0, \ldots, 0, u_j, 0, \ldots, 0) \text{ for some } j \in \{1, \ldots, n\},$$

where $(-\Delta)^m u_j \geq 0$ *in* $\mathbb{R}^N$.

Proof Part (i) follows immediately from Theorem 4.11. For part (ii), let $(u_1, \ldots, u_n)$ be a nontrivial nonnegative solution of (4.56). Assume, without loss of generality, that $u_1 \not\equiv 0$. Let $i \in \{2, \ldots, n\}$. Since $e_{1i} = e_{i1} = 1$, by Theorem 4.13 we see that $u_i \equiv 0$. This indicates that $u_i \equiv 0$ for each $i \in \{2, \ldots, n\}$. The conclusion holds. □

4.7   Conclusions and Further Remarks

We discussed in this chapter some partial differential inequalities driven by the polyharmonic operator $(-\Delta)^m$. One of the key tools we employ in this chapter is the integral representation of solutions as stated in Theorem 4.1 that is originally due to [CDM08]. The study of integral representations is a central topic in potential theory and goes back to the work of F. Riesz [R30] in 1930. The well-known Riesz representation theorem states that if $u \in L^1_{loc}(\mathbb{R}^N)$, $N \geq 3$ is nonnegative and satisfies

$$(-\Delta)u = \mu \quad \text{in } \mathscr{D}'(\mathbb{R}^N), \quad N > 2. \tag{4.72}$$

for some positive Radon measure μ, then u admits the representation

$$u(x) = \ell + c(N) \int_{\mathbb{R}^N} |x - y|^{2-N} d\mu(y) \quad \text{for a.a. } x \in \mathbb{R}^N, \tag{4.73}$$

where $\ell \geq 0$ is a suitable constant and $c(N) > 0$ is a universal constant. This implies that the nonnegative solution of (4.72) is unique up to a additive constant. Therefore, the following question arises:

which conditions assure that a solution u of (4.72) is nonnegative?

An answer given in [CDM08] is that for a suitable constant $\ell \geq 0$, u must satisfy the so-called ring condition (see (4.6), namely,

$$\liminf_{R \to \infty} \frac{1}{R^N} \int\limits_{R \leq |y-x| \leq 2R} |u(y) - \ell| dy = 0 \quad \text{for a.a. } x \in \mathbb{R}^N, \tag{R}$$

and hence the representation (4.73) holds (with the same ℓ). This sufficient condition (R) is also a necessary condition for the integral representation of the solutions to (4.72).

For the higher-order case, namely,

$$(-\Delta)^m u = \mu \quad \text{in } \mathscr{D}'(\mathbb{R}^N), \quad N > 2m, \quad m \geq 1, \tag{4.74}$$

the situation is compound. Clearly, in this case, the analog of (4.73) is given by

$$u(x) = \ell + c(N, m) \int_{\mathbb{R}^N} |x - y|^{2m-N} d\mu(y) \quad \text{for a.a. } x \in \mathbb{R}^N. \tag{4.75}$$

For $m \geq 2$, one cannot represent all the nonnegative solution of (4.74) by (4.75). Indeed, the function $u(x) = |x|^2$ is a nonnegative solution of (4.74) with $\mu \equiv 0$, which cannot be represented by (4.75). Therefore, the Riesz representation theorem does not hold verbatim. However, one of the main results in [CDM08] establishes that a solution u of (4.74) admits the integral representation (4.75) for some $\ell \in \mathbb{R}$ if and only if u satisfies the same ring condition (R). Another relevant statement in Theorem 4.1 is that a solution u of (4.74) can be represented by (4.75) if and only if it is superharmonic.

For a more general statement of the *Riesz representation theorem* for superharmonic functions defined on a Greenian open set $\Omega \subset \mathbb{R}^N$, see [AG01]. Such integral representations have been extended to the parabolic setting [Kem72, Wat76], to the framework of some degenerate elliptic operators [BLU07, DMP06] and adapted to the study of partial differential equations in a number of ways (see, e.g. [GMT11, GT17, Tal07]). An extension of Theorem 4.1 is recently discussed in [DG22] and concerns the integral representations of nonnegative solutions to $P(-\Delta)u = \mu$ in $\mathscr{D}'(\Omega)$, where $P(z)$ is a polynomial with real coefficients.

One major application of integral representations in partial differential equations is that they provide a powerful tool to discuss positivity of solutions (as already said) and Liouville-type theorems as well [CDM08, DG22, DMP06].

For instance, knowing that all the solutions of

$$(-\Delta)^m u = |u|^q \quad \text{in } \mathscr{D}'(\mathbb{R}^N), \ N > 2m, \ q > 1, \tag{4.76}$$

can be represented by

$$u(x) = c \int |x - y|^{2m-N} |u|^q(y) dy \quad \text{for a.a. } x \in \mathbb{R}^N$$

(see [CDM08, DG22]), we can deduce several properties of the solutions (such as existence, nonexistence, positivity, regularity and uniqueness) by invoking appropriate results available in the theory of integral equations (see, e.g. [Li04]).

Chapter 5
Quasilinear Parabolic Inequalities with Convolution Terms

5.1 Introduction

In this chapter, we deal with the nonexistence of nonnegative solutions of the following parabolic problems:

$$\begin{cases} \dfrac{\partial u}{\partial t} \pm \mathscr{L}u \geq (K * u^p)u^q & \text{in } \mathbb{R}^N \times (0, \infty), \ N \geq 1 \\[2mm] u(x, 0) = u_0(x) \geq 0 & \text{in } \mathbb{R}^N, \end{cases} \tag{5.1}$$

where $u_0 \in L^1_{loc}(\mathbb{R}^N)$, $u_0 \geq 0$ and the differential operator $\mathscr{L}u = \operatorname{div}\mathscr{A}(x, u, \nabla u)$ are weakly-m-coercive for some $m > 1$ as already stated in Definition 2.1. This means that $\mathscr{A} : \mathbb{R}^N \times \mathbb{R} \times \mathbb{R}^N \to \mathbb{R}^N$ is a Carathéodory function and there exists $c_0 > 0$ such that the inequality

$$\mathscr{A}(x, z, \xi) \cdot \xi \geq c_0 |\mathscr{A}(x, z, \xi)|^{m'} \tag{5.2}$$

holds for all $(x, z, \xi) \in \mathbb{R}^N \times [0, \infty) \times \mathbb{R}^N$ with $m' = m/(m - 1)$. The above inequality implies

$$\mathscr{A}(x, z, 0) = 0, \qquad \mathscr{A}(x, z, \xi) \cdot \xi \geq 0 \tag{5.3}$$

for every $(x, z, \xi) \in \mathbb{R}^N \times [0, \infty) \times \mathbb{R}^N$. As we have already seen in Chap. 2, the main prototypes for $\mathscr{L}$ are the m-Laplace operator and the m-mean curvature operator.

Furthermore, related to the nonlinearity on the right-hand side of (5.1), we assume $p, q > 0$. The function $K \in C(\mathbb{R}^N \setminus \{0\})$, $K > 0$ satisfies $\liminf_{x \to 0} K(x) > 0$ and there exists $\rho > 0$ and $0 < \alpha < m/2$ such that

© The Author(s), under exclusive license to Springer Nature Switzerland AG 2022
M. Ghergu, *Partial Differential Inequalities with Nonlinear Convolution Terms*, SpringerBriefs in Mathematics, https://doi.org/10.1007/978-3-031-21856-9_5

$$K(x) \geq c|x|^{-\alpha} \quad \text{for all } x \in \mathbb{R}^N, \ |x| > \rho, \tag{5.4}$$

where $c > 0$ is a positive constant. Also, by $K * u^p$ we denote the standard convolution operator defined by

$$(K * u^p)(x, t) = \int_{\mathbb{R}^N} K(x - y)u^p(y, t)dy \quad \text{for all } (x, t) \in \mathbb{R}^N \times (0, \infty).$$

We prefer to discuss separately the two problems defined by (5.1) according to the $\pm$ sign, as different classes of nonnegative solutions appear naturally in our approach.

5.2 The Inequality $\frac{\partial u}{\partial t} - \mathscr{L}u \geq (K * u^p)u^q$ in $\mathbb{R}^N \times (0, \infty)$

In this section, we discuss the problem

$$\begin{cases} \dfrac{\partial u}{\partial t} - \mathscr{L}u \geq (K * u^p)u^q & \text{in } \mathbb{R}^N \times (0, \infty), \ N \geq 1 \\[2mm] u(x, 0) = u_0(x) \geq 0 & \text{in } \mathbb{R}^N, \end{cases} \tag{5.5}$$

We are interested in *nonnegative weak solutions* of (5.5), that is, nonnegative functions $u(x, t)$, belonging to the class $\mathscr{S}$ given by those $u \in W_{\text{loc}}^{1,m}(\mathbb{R}^N \times (0, \infty))$ which fulfil the two conditions below:

(i) $\mathscr{A}(x, u, \nabla u) \in [L_{\text{loc}}^{m'}(\mathbb{R}^N \times (0, \infty))]^N$,
(ii) $(K * u^p)u^q \in L_{\text{loc}}^1(\mathbb{R}^N \times (0, \infty))$,

and such that for any nonnegative test function $\varphi \in C_c^1(\mathbb{R}^N \times \mathbb{R})$, we have

$$\int_0^\infty \int_{\mathbb{R}^N} (K * u^p)u^q \varphi dx dt \leq \int_0^\infty \int_{\mathbb{R}^N} \frac{\partial u}{\partial t}\varphi dx dt + \int_0^\infty \int_{\mathbb{R}^N} \mathscr{A}(x, u, \nabla u) \cdot \nabla \varphi dx dt \tag{5.6}$$

or equivalently

$$\int_0^\infty \int_{\mathbb{R}^N} (K * u^p)u^q \varphi \, dx \, dt \leq -\int_{\mathbb{R}^N} u_0(x)\varphi(x, 0) \, dx - \int_0^\infty \int_{\mathbb{R}^N} u \frac{\partial \varphi}{\partial t} \, dx \, dt$$
$$+ \int_0^\infty \int_{\mathbb{R}^N} \mathscr{A}(x, u, \nabla u) \cdot \nabla \varphi \, dx \, dt. \tag{5.7}$$

The main result concerning (5.5) is the following.

Theorem 5.1 *Assume*

$$0 < \alpha < m/2 \quad \text{and} \quad m > \frac{2N + 1}{N + 1}. \tag{5.8}$$

If

$$2\max\{1, m-1\} < p+q \leq m-1 + \frac{N-\alpha+m}{N+\alpha}, \qquad (5.9)$$

then problem (5.5) does not have nonnegative nontrivial solutions $u \in \mathscr{S}$.

Let us note that in particular, condition (5.9) yields the following upper bounds for α:

$$\alpha < \begin{cases} 1 - \dfrac{m-2}{m}N & \text{if } 2 < m < \frac{2N}{N-1}, \\[2mm] \dfrac{m - (2-m)N}{4 - m} & \text{if } \frac{2N+1}{N+1} < m < 2. \end{cases} \qquad (5.10)$$

Theorem 5.1 applies to the particular case where $\mathscr{L}$ is the m-Laplace operator or the m-mean curvature operator. We obtain directly the following results.

Corollary 5.2 *Assume*

$$0 < \alpha < 1 \quad and \quad 2 < p+q \leq 2 + \frac{2(1-\alpha)}{N+\alpha}.$$

Then, the problem

$$\begin{cases} \dfrac{\partial u}{\partial t} - \operatorname{div}\left(\dfrac{\nabla u}{\sqrt{1 + |\nabla u|^2}}\right) \geq (K * u^p)u^q & \text{in } \mathbb{R}^N \times (0, \infty), \\[4mm] u(x, 0) = u_0(x) \geq 0 & \text{in } \mathbb{R}^N, \end{cases}$$

does not have nonnegative nontrivial solutions.

Corollary 5.3 *Assume*

$$0 < \alpha < m/2 \quad and \quad m > \frac{2N+1}{N+1}\,(\in (1, 2)).$$

If (5.9) holds, then the problem

$$\begin{cases} \dfrac{\partial u}{\partial t} - \Delta_m u \geq (K * u^p)u^q & \text{in } \mathbb{R}^N \times (0, \infty), \\[3mm] u(x, 0) = u_0(x) \geq 0 & \text{in } \mathbb{R}^N, \end{cases}$$

does not have nonnegative nontrivial solutions.

Proof Suppose by contradiction that (5.5) admits a nonnegative nontrivial solution $u \in \mathscr{S}$. We start with the following result which provides extra local integrability of u.   □

Lemma 5.4 *Let $u \in \mathscr{S}$ be a nonnegative solution of (5.5). Then,*

$$u^{(p+q)/2} \in L^1_{loc}(\mathbb{R}^N \times [0, \infty)). \tag{5.11}$$

Proof Let $R > \rho$ be large, where $\rho > 0$ appears in (5.4). For $x \in B_R(0)$, using (5.4) we have

$$(K*u^p)(x, t) \geq \int_{\mathbb{B}_R} K(x - y)u^p(y, t)dy$$

$$= \int_{|x-y|\leq\rho} K(x - y)u^p(y, t))dy + \int_{\substack{|x-y|>\rho \\ |x|, |y|<R}} K(x - y)u^p(y, t)dy$$

$$\geq \inf_{z\in B_1(0)} K(z) \int_{|x-y|\leq\rho} u^p(y, t)dy + c \int_{\substack{|x-y|>\rho \\ |x|, |y|<R}} |x - y|^{-\alpha}u^p(y, t)dy$$

$$\geq \inf_{z\in B_1(0)} K(z) \int_{|x-y|\leq\rho} u^p(y, t)dy + c(2R)^{-\alpha} \int_{|x-y|>\rho, |y|<R} u^p(y, t)dy$$

$$\geq CR^{-\alpha} \left\{ \int_{|x-y|\leq\rho, |y|<R} u^p(y, t)dy + \int_{|x-y|>\rho, |y|<R} u^p(y, t)dy \right\}$$

$$\geq CR^{-\alpha} \int_{B_R(0)} u^p(y, t)dy, \tag{5.12}$$

provided $R > \rho$ is large enough.

Next, using (5.12) and Hölder's inequality, we find

$$\infty > \int_0^T \int_{B_R(0)} (K * u)(x, t)u^q(x, t)dxdt$$

$$\geq CR^{-\alpha} \int_0^T \left(\int_{B_R(0)} u^p(x)dx \right)\left(\int_{B_R(0)} u^q(x)dx \right)dt \qquad \text{(by (5.12))}$$

$$\geq CR^{-\alpha} \int_0^T \left(\int_{B_R(0)} u^{(p+q)/2}(x)dx \right)^2 dt \qquad \text{(by Hölder's inequality on $B_R(0)$)}$$

$$\geq \frac{CR^{-\alpha}}{T} \left(\int_0^T \int_{B_R(0)} u^{(p+q)/2}(x)dx\, dt \right)^2 \qquad \text{(by Hölder's inequality on $[0, T]$)}$$

which shows that $u^{(p+q)/2} \in L^1(B_R(0) \times [0, T])$ and concludes our proof.   □

From (5.9) we have $(p + q)/2 > \max\{1, m - 1\}$. We may thus choose $d > 0$ small enough such that $(p + q)/2 > (m - 1)(d + 1)$. Using Hölder's inequality, we deduce

$$u^{1-d}, \ u^{(m-1)(1-d)}, \ u^{(m-1)(d+1)} \in L^1_{loc}(\mathbb{R}^N \times [0, \infty)). \tag{5.13}$$

This will ensure that all integrals in this section are finite.

The proof of Theorem 5.1 will be achieved along several lemmas. First we want to precise the choice of the test functions in (5.6). Take a standard cutoff function $\xi \in C^1[0, \infty)$ such that

- $\xi = 1$ in $(0, 1)$, $\xi = 0$ in $(2, \infty)$;
- $0 \leq \xi \leq 1$ and $|\xi'| \leq C$ in $[0, \infty)$, for some $C > 0$.

Now take $R > 0$ and consider the functions

$$\chi_R(x) = \xi\left(\frac{|x|}{R}\right), \quad \eta_R(t) = \xi\left(\frac{t}{R^\gamma}\right), \tag{5.14}$$

with $\gamma \geq 1$. Clearly $\chi_R(x) = 1$ in $B_R(0)$, where $B_R(0)$ denotes the open ball in $\mathbb{R}^N$, centred at the origin and having radius $R > 0$.

Finally define, for all $R > 0$, the nonnegative cutoff function

$$\psi : \mathbb{R}^N \times [0, \infty) \to [0, \infty), \quad \psi(x, t) = \chi_R(x)\,\eta_R(t). \tag{5.15}$$

Note that ψ is the restriction to $\mathbb{R}^N \times [0, \infty)$ of a $C^1_c(\mathbb{R}^N \times \mathbb{R})$ function, and by the shape of ψ in (5.15), the following inequalities hold for $\varsigma > 1$, $\gamma \geq 1$, k and R sufficiently large:

$$\int_0^\infty \int_{\mathbb{R}^N} \psi^{k-\varsigma} |\nabla\psi|^\varsigma dx\,dt \leq cR^{-\varsigma+N+\gamma}, \tag{5.16}$$

$$\int_0^\infty \int_{\mathbb{R}^N} \psi^{k-\varsigma} \left|\frac{\partial\psi}{\partial t}\right|^\varsigma dx\,dt \leq cR^{-\gamma\varsigma+N+\gamma}, \tag{5.17}$$

where $c > 0$ is a constant.

Lemma 5.5 *Let $u \in \mathscr{S}$ be a nonnegative solution of (5.5) that satisfies (5.13) and let ψ be defined by (5.15). Then,*

$$\int_0^\infty \int_{\mathbb{R}^N} (K * u^p)u^{q-d}\psi^k \, dx\,dt + \int_0^\infty \int_{\mathbb{R}^N} u^{-d-1}\psi^k |\mathscr{A}(x, u, \nabla u)|^{m'} dx\,dt$$

$$\leq c_1 \int_0^\infty \int_{\mathbb{R}^N} u^{1-d}\psi^{k-1}\left|\frac{\partial\psi}{\partial t}\right| dx\,dt + c_2 \int_0^\infty \int_{\mathbb{R}^N} u^{m-d-1}\psi^{k-m}|\nabla\psi|^m dx\,dt, \tag{5.18}$$

for some constants $c_1, c_2 > 0$.

Proof Let $\varepsilon > 0$ be sufficiently small and let $\{\xi_\varepsilon\}_{\varepsilon>0}$ be a standard family of mollifiers. For $\tau > 0$ we define

$$\tilde{u}_\varepsilon(x, t) := \tau + \int_{\mathbb{R}^N} \xi_\varepsilon(x - y, t)u(y, t)\, dy \quad \text{and} \quad u_\tau(x, t) := \tau + u(x, t)$$

for $(x, t) \in \mathbb{R}^N \times (0, \infty)$. Clearly $\tilde{u}_\varepsilon, u_\tau \geq \tau > 0$. In particular, since $u \in L^1_{\text{loc}}(\mathbb{R}^N \times (0, \infty))$, we have $\tilde{u}_\varepsilon \in C^1(\mathbb{R}^N \times (0, \infty))$. Let $\psi(x, t)$ be defined in (5.15). We take in the weak formulation (5.6) the test function

$$\varphi(x, t) = \tilde{u}_\varepsilon^{-d}\, \psi^k(x, t) \geq 0,$$

where $k > 0$. Thus,

$$\int_0^\infty \int_{\mathbb{R}^N} (K * u^p)u^q \tilde{u}_\varepsilon^{-d} \psi^k\, dx\, dt + d\int_0^\infty \int_{\mathbb{R}^N} \psi^k \tilde{u}_\varepsilon^{-d-1} \mathscr{A}(x, u, \nabla u) \cdot \nabla \tilde{u}_\varepsilon\, dx\, dt$$

$$\leq \int_0^\infty \int_{\mathbb{R}^N} \frac{\partial u}{\partial t} \tilde{u}_\varepsilon^{-d}\, \psi^k\, dx\, dt + k\int_0^\infty \int_{\mathbb{R}^N} \psi^{k-1} \tilde{u}_\varepsilon^{-d} \mathscr{A}(x, u, \nabla u) \cdot \nabla \psi\, dx\, dt.$$

Since $\tilde{u}_\varepsilon \to u_\tau$ in $W^1_{\text{loc}}(\mathbb{R}^N)$ as $\varepsilon \to 0$, using Lebesgue dominated convergence theorem, and using $\nabla u_\tau = \nabla u$, we arrive at

$$\int_0^\infty \int_{\mathbb{R}^N} (K * u^p)u^q u_\tau^{-d} \psi^k\, dx\, dt + d\int_0^\infty \int_{\mathbb{R}^N} \psi^k u_\tau^{-d-1} \mathscr{A}(x, u, \nabla u) \cdot \nabla u\, dx\, dt$$

$$\leq \int_0^\infty \int_{\mathbb{R}^N} \frac{\partial u}{\partial t} u_\tau^{-d} \psi^k\, dx\, dt + k\int_0^\infty \int_{\mathbb{R}^N} \psi^{k-1} u_\tau^{-d} |\mathscr{A}(x, u, \nabla u)||\nabla \psi|\, dx\, dt. \tag{5.19}$$

Using the weak m-coerciveness of $\mathscr{A}$, from (5.2) we deduce

$$\int_0^\infty \int_{\mathbb{R}^N} (K * u^p)u^q u_\tau^{-d} \psi^k\, dx\, dt + dc_0\int_0^\infty \int_{\mathbb{R}^N} \psi^k u_\tau^{-d-1} |\mathscr{A}(x, u, \nabla u)|^{m'}\, dx\, dt$$

$$\leq \int_0^\infty \int_{\mathbb{R}^N} \frac{\partial u}{\partial t} u_\tau^{-d} \psi^k\, dx\, dt + k\int_0^\infty \int_{\mathbb{R}^N} \psi^{k-1} u_\tau^{-d} |\mathscr{A}(x, u, \nabla u)||\nabla \psi|\, dx\, dt. \tag{5.20}$$

Now consider the first term on the right-hand side of (5.20) and we claim that

$$\int_0^\infty \int_{\mathbb{R}^N} \frac{\partial u}{\partial t} u_\tau^{-d}\, \psi^k\, dx\, dt \leq \frac{k}{1 - d}\int_0^\infty \int_{\mathbb{R}^N} u_\tau^{1-d}\psi^{k-1} \left|\frac{\partial \psi}{\partial t}\right|\, dx\, dt. \tag{5.21}$$

Indeed, since $\frac{\partial u_\tau}{\partial t} = \frac{\partial u}{\partial t}$, by definition of u_τ, we have

$$\int_0^\infty \int_{\mathbb{R}^N} \frac{\partial u}{\partial t} u_\tau^{-d} \psi^k \, dx \, dt = \int_0^\infty \int_{\mathbb{R}^N} \frac{\partial u_\tau}{\partial t} u_\tau^{-d} \, \psi^k \, dx \, dt$$

$$= \frac{1}{1-d} \int_0^\infty \int_{\mathbb{R}^N} \frac{\partial u_\tau^{1-d}}{\partial t} \psi^k \, dx \, dt$$

$$= \frac{1}{1-d} \int_0^\infty \int_{\mathbb{R}^N} \left[\frac{\partial (u_\tau^{1-d} \psi^k)}{\partial t} - u_\tau^{1-d} \frac{\partial \psi^k}{\partial t} \right] dx \, dt$$

$$= -\frac{1}{1-d} \int_{\mathbb{R}^N} u_\tau^{1-d}(x, 0) \psi^k(x, 0) \, dx - \frac{k}{1-d} \int_0^\infty \int_{\mathbb{R}^N} u_\tau^{1-d} \, \psi^{k-1} \frac{\partial \psi}{\partial t} \, dx \, dt,$$

where the last equality is due to $\psi \in C_c^1(\mathbb{R}^N \times [0, \infty))$; hence, $\lim_{t\to\infty} \psi(x, t) = 0$. Consequently, (5.21) follows immediately from $u_\tau(x, 0) = u_0(x) + \tau > 0$, $\psi \geq 0$ and $0 < d < 1$.

Further, by Young's inequality , we have

$$u_\tau^{-d} \psi^{k-1} |\mathscr{A}(x, u, \nabla u)| \|\nabla \psi\| \leq \frac{dc_0}{2k} u_\tau^{-d-1} \psi^k |\mathscr{A}(x, u, \nabla u)|^{m'} + C u_\tau^{m-d-1} \psi^{k-m} |\nabla \psi|^m.$$

Thanks to property (i) in the definition of $\mathscr{S}$ and the fact that $u_\tau \geq \tau > 0$ and $0 \leq \psi \leq 1$, we have

$$u_\tau^{-d-1} \, \psi^k |\mathscr{A}(x, u, \nabla u)|^{m'} \in L_{loc}^1(\mathbb{R}^N \times [0, \infty)).$$

Thus, a combination of (5.20) and (5.21) yields

$$\int_0^\infty \int_{\mathbb{R}^N} (K * u^p) u^q u_\tau^{-d} \psi^k \, dx \, dt + \int_0^\infty \int_{\mathbb{R}^N} u_\tau^{-d-1} \psi^k |\mathscr{A}(x, u, \nabla u)|^{m'} dx \, dt$$

$$\leq c_1 \int_0^\infty \int_{\mathbb{R}^N} u_\tau^{1-d} \psi^{k-1} \left| \frac{\partial \psi}{\partial t} \right| dx \, dt + c_2 \int_0^\infty \int_{\mathbb{R}^N} u_t^{m-d-1} \psi^{k-m} |\nabla \psi|^m dx \, dt,$$

with $c_1, c_2 > 0$. Since all the exponents of u on the right-hand side are positive, we let $\tau \to 0$ and apply Fatou and Lebesgue theorems, to obtain (5.18). $\square$

Lemma 5.6 *Let*

$$\ell > \max\{1, m - 1\} \tag{5.22}$$

and let $u \geq 0$ be a solution of (5.5) such that $u^\ell \in L_{loc}^1(\mathbb{R}^N \times [0, \infty))$. Define

$$J(t) = \int_{\mathbb{R}^N} u^\ell(x, t) \psi^k(x, t) dx, \tag{5.23}$$

where ψ is given by (5.15). Then,

$$\int_0^\infty \int_{\mathbb{R}^N} (K * u^p) u^q \psi^k \, dx \, dt \leq c_1 \left(\int_0^{2R^\gamma} J(t) \right)^{1/\ell} \cdot R^{\frac{N+\gamma}{\ell} - \gamma}$$

$$+ c_2 \left(\int_0^{2R^\gamma} J(t) \right)^{\frac{2(m-1)}{m\ell}} \cdot R^{(N+\gamma)\left(1 - \frac{2}{m'\ell}\right) - 1 - \frac{\gamma}{m'}} \tag{5.24}$$

$$+ c_3 \left(\int_0^{2R^\gamma} J(t) \right)^{\frac{m-1}{\ell}} \cdot R^{(N+\gamma)\left(1 - \frac{m-1}{\ell}\right) - m},$$

for some constants $c_1, c_2, c_3 > 0$.

Proof We first choose $\varphi = \psi^k$ in the weak formulation (5.7), with ψ given by (5.15). Since $u_0, \varphi \geq 0$, we find

$$\int_0^\infty \int_{\mathbb{R}^N} (K * u^p) u^q \psi^k \, dx \, dt \leq -\int_0^\infty \int_{\mathbb{R}^N} u \, (\psi^k)_t \, dx \, dt$$

$$+ \iint_{\text{supp}(\nabla\psi)} |\mathscr{A}(x, u, \nabla u)| |\nabla \psi^k| dx \, dt. \tag{5.25}$$

From now on, the constants c_1, c_2 and c_3 will assume different values.

By Hölder's inequality, we obtain

$$\int_0^\infty \int_{\mathbb{R}^N} (K * u^p) u^q \psi^k \, dx \, dt \leq C \left(\iint_{\text{supp}(\psi_t)} u^\ell \psi^k dx dt \right)^{1/\ell} \cdot \left(\iint_{\text{supp}(\psi_t)} \psi^{k - \ell'} |\psi_t|^{\ell'} \right)^{1/\ell'}$$

$$+ \iint_{\text{supp}(\nabla\psi)} |\mathscr{A}(x, u, \nabla u)| |\nabla \psi^k| dx \, dt.$$

$$\tag{5.26}$$

We next estimate the last integral in (5.26). First, by Hölder's inequality, we have

$$\iint_{\text{supp}(\nabla\psi)} |\mathscr{A}(x, u, \nabla u)| |\nabla \psi^k| dx \, dt \leq k \left(\int_0^\infty \int_{\mathbb{R}^N} u^{-d-1} \psi^k |\mathscr{A}(x, u, \nabla u)|^{m'} dx \, dt \right)^{1/m'}$$

$$\cdot \left(\int_0^\infty \int_{\mathbb{R}^N} u^{(d+1)(m-1)} \psi^{k-m} |\nabla \psi|^m dx \, dt \right)^{1/m}.$$

Next, we use the estimate (5.18) from Lemma 5.5 and the standard inequality

$$(a + b)^r \leq c(a^r + b^r) \quad \text{for all} \quad a, b, r > 0.$$

We deduce

$$\int_0^\infty \int_{\mathbb{R}^N} |\mathscr{A}(x, u, \nabla u)||\nabla \psi^k|dx\,dt \leq c_1 \left(\int_0^\infty \int_{\mathbb{R}^N} u^{1-d}\psi^{k-1}|\psi_t|\,dxdt \right)^{1/m'}$$

$$\cdot \left(\int_0^\infty \int_{\mathbb{R}^N} u^{(d+1)(m-1)}\psi^{k-m}|\nabla \psi|^m dx\,dt \right)^{1/m}$$

$$+ c_2 \left(\int_0^\infty \int_{\mathbb{R}^N} u^{m-d-1}\psi^{k-m}|\nabla \psi|^m dx\,dt \right)^{1/m'}$$

$$\cdot \left(\int_0^\infty \int_{\mathbb{R}^N} u^{(d+1)(m-1)}\psi^{k-m}|\nabla \psi|^m dx\,dt \right)^{1/m},$$

$$(5.27)$$

where $0 < d < \min\{1, m-1\}$. We now use Hölder's inequality in all the factors of the right-hand side, so that

$$\int_0^\infty \int_{\mathbb{R}^N} u^{1-d}\psi^{k-1}|\psi_t|\,dxdt \leq \left(\iint_{\text{supp}(\psi_t)} u^\ell \psi^k \right)^{1/\sigma} \left(\iint_{\text{supp}(\psi_t)} \psi^{k-\sigma'}|\psi_t|^{\sigma'} \right)^{1/\sigma'},$$

$$(5.28)$$

where

$$\sigma = \frac{\ell}{1-d}, \qquad \sigma' = \frac{\ell}{\ell+d-1};$$

$$\int_0^\infty \int_{\mathbb{R}^N} u^{(d+1)(m-1)}\psi^{k-m}|\nabla \psi|^m dx\,dt$$

$$\leq \left(\iint_{\text{supp}(\nabla \psi)} u^\ell \psi^k \right)^{1/\eta} \left(\int_0^\infty \int_{\mathbb{R}^N} \psi^{k-m\eta'}|\nabla \psi|^{m\eta'} \right)^{1/\eta'},$$

$$(5.29)$$

where

$$\eta = \frac{\ell}{(d+1)(m-1)}, \qquad \eta' = \frac{\ell}{\ell - m + 1 - d(m-1)}; \qquad (5.30)$$

$$\int_0^\infty \int_{\mathbb{R}^N} u^{m-1-d}\psi^{k-m}|\nabla \psi|^m \leq \left(\iint_{\text{supp}(\nabla \psi)} u^\ell \psi^k \right)^{1/\theta} \left(\int_0^\infty \int_{\mathbb{R}^N} \psi^{k-m\theta'}|\nabla \psi|^{m\theta'} \right)^{1/\theta'}$$

$$(5.31)$$

where

$$\theta = \frac{\ell}{m-d-1}, \qquad \theta' = \frac{\ell}{\ell - m + d + 1}. \qquad (5.32)$$

We next replace (5.28), (5.29) and (5.31) in (5.27). With $J(t)$ defined in (5.23), we obtain

$$\int_0^\infty \int_{\mathbb{R}^N} |\mathscr{A}(x, u, \nabla u)||\nabla \psi^k| dx\, dt \leq$$

$$c_1 \left(\int_0^\infty J(t) \right)^{\frac{1}{m'\sigma} + \frac{1}{m\eta}} \left(\int_0^\infty \int_{\mathbb{R}^N} \psi^{k - \sigma'} |\psi_t|^{\sigma'} \right)^{1/m'\sigma'} \left(\int_{\mathbb{R}^N} \psi^{k - m\eta'} |\nabla \psi|^{m\eta'} \right)^{1/m\eta'}$$

$$+ c_2 \left(\int_0^\infty J(t) \right)^{\frac{1}{m'\theta} + \frac{1}{m\eta}} \left(\int_0^\infty \int_{\mathbb{R}^N} \psi^{k - m\theta'} |\nabla \psi|^{m\theta'} \right)^{1/m'\theta'} \left(\int_0^\infty \int_{\mathbb{R}^N} \psi^{k - m\eta'} |\nabla \psi|^{m\eta'} \right)^{1/m\eta'}.$$

Inserting the above inequality in (5.26) and using (5.16) and (5.17) together with the fact that $J(t) = 0$ for $t \geq 2R^\gamma$ from the definition of the test function ψ, we find

$$\int_0^\infty \int_{\mathbb{R}^N} (K * u^p) u^q \psi^k \, dx\, dt \leq c_1 \left(\int_0^{2R^\gamma} J(t) \right)^{1/\ell} \cdot R^{\frac{N+\gamma-\gamma\ell'}{\ell'}}$$

$$+ c_2 \left(\int_0^{2R^\gamma} J(t) \right)^{\frac{1}{m'\sigma} + \frac{1}{m\eta}} \cdot R^{\frac{N+\gamma-\gamma\sigma'}{m'\sigma'} + \frac{N+\gamma-m\eta'}{m\eta'}}$$

$$+ c_3 \left(\int_0^{2R^\gamma} J(t) \right)^{\frac{1}{m'\theta} + \frac{1}{m\eta}} \cdot R^{\frac{N+\gamma-m\theta'}{m'\theta'} + \frac{N+\gamma-m\eta'}{m\eta'}}$$

$$= c_1 \left(\int_0^{2R^\gamma} J(t) \right)^{1/\ell} \cdot R^{\frac{N+\gamma-\gamma\ell'}{\ell'}}$$

$$+ c_2 \left(\int_0^{2R^\gamma} J(t) \right)^{\frac{1}{m'\sigma} + \frac{1}{m\eta}} \cdot R^{(N+\gamma)\left(\frac{1}{m'\sigma'} + \frac{1}{m\eta'}\right) - 1 - \frac{\gamma}{m'}}$$

$$+ c_3 \left(\int_0^{2R^\gamma} J(t) \right)^{\frac{1}{m'\theta} + \frac{1}{m\eta}} \cdot R^{(N+\gamma)\left(\frac{1}{m'\theta'} + \frac{1}{m\eta'}\right) - m}. \tag{5.33}$$

Consequently,

$$\frac{1}{m'\sigma} + \frac{1}{m\eta} = \frac{2(m-1)}{m\ell}, \qquad \frac{1}{m'\theta} + \frac{1}{m\eta} = \frac{m-1}{\ell},$$

$$\frac{1}{m'\sigma'} + \frac{1}{m\eta'} = \frac{m\ell - 2(m-1)}{m\ell}, \qquad \frac{1}{m'\theta'} + \frac{1}{m\eta'} = \frac{\ell - m + 1}{\ell}.$$

Observe that all quantities in the above expressions are positive. Indeed, by (5.22) we deduce that $\ell \geq 2(m-1)/m$. Using (5.33) and the values of the parameters involved, we get the estimate (5.24). $\qquad \square$

Lemma 5.7 *Let $u \geq 0$ be a solution of (5.5) and J be given by (5.23) with $\ell = (p+q)/2 > 1$. Then*

$$\int_0^{2R^\gamma} J(t)^2 dt \leq c\left(R^{\beta_1} + R^{\beta_2} + R^{\beta_3}\right), \tag{5.34}$$

where $c > 0$ is a constant and

$$\beta_1 = \frac{(N + \alpha)(p + q) - 2N - \gamma}{p + q - 1},$$

$$\beta_2 = \frac{(N + \alpha - 1)m(p + q) - 4N(m - 1) + \gamma[p + q - 2(m - 1)]}{m(p + q) - 2(m - 1)},$$

$$\beta_3 = \frac{(N + \alpha - m)(p + q) - 2N(m - 1) + \gamma(p + q - m + 1)}{p + q - m + 1}.$$

Proof First note that by (5.11), we can choose $\ell = (p + q)/2$ which by (5.9) satisfies (5.22) so that the requirement of Lemma 5.6 is satisfied. Now, observe that $\mathrm{supp}\,(\psi^k) \subset B_{2R}(0) \times [0, 2R^\gamma)$. If $(x, t) \in B_{2R}(0) \times [0, \infty)$, then, in the same way as we estimated (5.12), we find

$$(K * u^p)(x, t) \geq CR^{-\alpha} \int_{B_{2R}(0)} u^p(y, t)dy \geq CR^{-\alpha} \int_{\mathbb{R}^N} u^p(y, t)\psi^k(y, t)dy,$$

$$(5.35)$$

provided $R > \rho$ is large enough.

Furthermore, for $\ell = (p + q)/2 > 1$, by Hölder's inequality, we have

$$\left(\iint_{\mathbb{R}^N \times \mathbb{R}^N} u^p(y, t)\psi^k(y, t)u^q(x, t)\psi^k(x, t)\,dx\,dy \right)^2$$

$$= \left(\iint_{\mathbb{R}^N \times \mathbb{R}^N} u^p(y, t)\psi^k(y, t)u^q(x, t)\psi^k(x, t)\,dx\,dy \right)$$

$$\cdot \left(\iint_{\mathbb{R}^N \times \mathbb{R}^N} u^p(x, t)\psi^k(x, t)u^q(y, t)\psi^k(y, t)\,dx\,dy \right)$$

$$\geq \left(\iint_{\mathbb{R}^N \times \mathbb{R}^N} u^{\frac{p+q}{2}}(x, t)u^{\frac{p+q}{2}}(y, t)\psi^k(x, t)\psi^k(y, t)\,dx\,dy \right)^2$$

$$= \left(\int_{\mathbb{R}^N} u^\ell(x, t)\psi^k(x, t)\,dx \right)^4 = J(t)^4,$$

where J is given by (5.23) with $\ell = (p + q)/2$. Hence,

$$\iint_{\mathbb{R}^N \times \mathbb{R}^N} u^p(y, t)\psi^k(y, t)u^q(x, t)\psi^k(x, t)\,dx\,dy \geq J^2(t).$$

Hence, using the above estimate and (5.35), we deduce

$$\int_{\mathbb{R}^N} (K * u^p)u^q \psi^k dx dt \geq CR^{-\alpha} \iint_{\mathbb{R}^N \times \mathbb{R}^N} u^p(y, t)\psi^k(y, t)u^q(x, t)\psi^k(x, t)dxdy$$

$$\leq CR^{-\alpha} J(t)^2.$$

$$(5.36)$$

As observed before inequality (5.33), we have $J(t) = 0$ if $t \geq 2R^\gamma$. Thus, (5.36) and inequality (5.24) in Lemma 5.6 yield

$$
\int_0^{2R^\gamma} J(t)^2 dt \leq c_1 \left(\int_0^{2R^\gamma} J(t)dt \right)^{1/\ell} \cdot R^{\frac{N+\gamma}{\ell'}-\gamma+\alpha}
$$

$$
+ c_2 \left(\int_0^{2R^\gamma} J(t)dt \right)^{\frac{2(m-1)}{m\ell}} \cdot R^{(N+\gamma)\left(1-\frac{2}{m'\ell}\right)-1-\frac{\gamma}{m'}+\alpha} \qquad (5.37)
$$

$$
+ c_3 \left(\int_0^{2R^\gamma} J(t)dt \right)^{\frac{m-1}{\ell}} \cdot R^{(N+\gamma)\left(1-\frac{m-1}{\ell}\right)-m+\alpha}.
$$

In addition, by Hölder's inequality , we find

$$
\int_0^{2R^\gamma} J(t)dt \leq \sqrt{2}R^{\gamma/2}\left(\int_0^{2R^\gamma} J(t)^2 dt \right)^{1/2}, \qquad (5.38)
$$

so that inequality (5.37) implies

$$
\int_0^{2R^\gamma} J(t)^2 dt \leq c_1 \left(\int_0^{2R^\gamma} J(t)^2 dt \right)^{1/2\ell} \cdot R^{\frac{N+\gamma}{\ell'}-\gamma+\alpha+\frac{\gamma}{2\ell}}
$$

$$
+ c_2 \left(\int_0^{2R^\gamma} J(t)^2 dt \right)^{\frac{m-1}{m\ell}} \cdot R^{(N+\gamma)\left(1-\frac{2}{m'\ell}\right)-1-\frac{\gamma}{m'}+\alpha+\frac{\gamma}{m'\ell}}
$$

$$
+ c_3 \left(\int_0^{2R^\gamma} J(t)^2 dt \right)^{\frac{m-1}{2\ell}} \cdot R^{(N+\gamma)\left(1-\frac{m-1}{\ell}\right)-m+\alpha+\frac{\gamma(m-1)}{2\ell}}.
$$

$$(5.39)$$

Now, applying Young's inequality , we arrive to

$$
\int_0^{2R^\gamma} J(t)^2 dt \leq \frac{1}{6} \int_0^{2R^\gamma} J(t)^2 dt + c_4 R^{\left[\frac{N+\gamma}{\ell'}+\alpha-\frac{\gamma}{(2\ell)'}\right](2\ell)'}
$$

$$
+ \frac{1}{6} \int_0^{2R^\gamma} J(t)^2 dt + c_5 R^{\left[(N+\gamma)\left(1-\frac{2}{m'\ell}\right)-1-\frac{\gamma}{m'\ell'}+\alpha\right]\left(\frac{m\ell}{m-1}\right)'}
$$

$$
+ \frac{1}{6} \int_0^{2R^\gamma} J(t)^2 dt + c_6 R^{\left[(N+\gamma)\left(1-\frac{m-1}{\ell}\right)-m+\alpha+\frac{\gamma(m-1)}{2\ell}\right]\left(\frac{2\ell}{m-1}\right)'}
$$

$$(5.40)$$

which yields (5.34), since

$$
\ell = \frac{p+q}{2}, \quad \left(\frac{m\ell}{m-1} \right)' = \frac{m(p+q)}{m(p+q)-2(m-1)}, \quad \left(\frac{2\ell}{m-1} \right)' = \frac{p+q}{p+q-m+1}.
$$

$$\square$$

Proof of Theorem 5.1 The proof will be carried out by taking a specific value of $\gamma \geq 1$ in the definition of the cutoff functions in (5.14). To this aim, we write β_1, β_2 and β_3 so that

$$\beta_1 = \frac{\Upsilon}{p+q-1}, \quad \beta_2 = \frac{2(m-1)}{m(p+q)-2(m-1)}\Upsilon, \quad \beta_3 = \frac{m-1}{p+q-m+1}\Upsilon$$

for a certain real constant Υ to be determined. Since

$$\beta_2 = \frac{2(m-1)}{m(p+q)-2(m-1)}\left[\frac{N+\alpha-1}{2(m-1)}m(p+q)-2N+\gamma\frac{p+q-2(m-1)}{2(m-1)}\right],$$

$$\beta_3 = \frac{m-1}{p+q-m+1}\left[\frac{N+\alpha-m}{m-1}(p+q)-2N+\gamma\frac{p+q-m+1}{m-1}\right],$$

we obtain that

$$\gamma = (m-2)(N+\alpha)+m, \quad \Upsilon = (N+\alpha)(p+q)-[Nm+\alpha(m-2)+m]. \quad (5.41)$$

Consequently, for

$$p+q < m-1+\frac{N-\alpha+m}{N+\alpha}$$

we obtain $\Upsilon < 0$, that is, β_1, β_2, $\beta_3 < 0$. By letting $R \to \infty$ in (5.34), it follows that

$$\int_0^\infty J(t)^2 dt = 0. \quad (5.42)$$

In turn, using the definition of J in (5.23) and the fact that $\psi \equiv 1$ in $B_R \times [0, R^\gamma)$, this easily yields $u \equiv 0$, a contradiction.

If

$$p+q = m-1+\frac{N-\alpha+m}{N+\alpha}, \quad (5.43)$$

then inequality (5.34) for $R \to \infty$ implies $J \in L^2(0, \infty)$.

Note that

$$\text{supp}\,(\nabla\psi) = (B_{2R} \setminus B_R) \times (0, 2R^\gamma),$$

$$\text{supp}\,\left(\frac{\partial\psi}{\partial t}\right) = B_{2R} \times (R^\gamma, 2R^\gamma).$$

Define next

$$\Theta_R = \max\left\{\int_{R^\gamma}^{2R^\gamma}\left(\int_{B_{2R}}u^\ell\psi^k dx\right)^2 dt\,,\,\int_0^{2R^\gamma}\left(\int_{B_{2R}\setminus B_R}u^\ell\psi^k dx\right)^2 dt\right\}.$$

Since $J \in L^2(0, \infty)$, it follows that $\Theta_R \to 0$ as $R \to \infty$. We now retake the estimate (5.26) and all the calculations that follow up to (5.39) to derive

$$\int_0^{2R^\gamma} J(t)^2 dt \le C\left\{\Theta_R^{\frac{1}{2\ell}} + \Theta_R^{\frac{m-1}{m\ell}} + \Theta_R^{\frac{m-1}{\ell}}\right\},$$

Now, letting $R \to \infty$ in the above estimate, it follows that $J \equiv 0$ and then $u \equiv 0$ which again contradicts our assumption.   □

We may slightly extend the result in Theorem 5.1 as follows:

Corollary 5.8 *Assume that $u_0 \in L^1_{loc}(\mathbb{R}^N)$ satisfies*

$$\int_{B_R} u_0(x)dx \ge cR^\nu \quad \text{for all } R > 0 \text{ large} \tag{5.44}$$

where $c > 0$ is a constant and the exponent ν fulfils

$$0 \le \nu < N + \alpha \quad \text{and} \quad 0 < \alpha < (m + \nu)/2.$$

If

$$2\max\{1, m - 1\} < p + q < m - 1 + \frac{N - \alpha + m}{N + \alpha - \nu}, \tag{5.45}$$

then problem (5.5) does not have nonnegative nontrivial solutions $u \in \mathscr{S}$ (where $\mathscr{S}$ is defined in Sect. 5.2).

In this case, condition (5.45) yields the following upper bound for α:

$$\alpha < \begin{cases} 1 - \dfrac{(m - 2)N - (m - 1)\nu}{m} & \text{if } 2 < m < \frac{2N}{N-1}, \\[3mm] \dfrac{m - (2 - m)N + (3 - m)\nu}{4 - m} & \text{if } \frac{2N+1}{N+1} < m < 2. \end{cases} \tag{5.46}$$

In particular, when $m > 2$, if we restrict the range of ν to the set

$$\frac{N(m - 2) - m}{m - 1} < \nu < N + \alpha,$$

then the upper bound $2N/(N - 1)$ on m in (5.46) can be removed.

Proof We choose in Lemma 5.7 the value of γ as follows:

$$\gamma = \frac{(N - \alpha)(m - 2) + m(p + q - 1)}{p + q - m + 1}.$$

Then

$$\beta_1 = \beta_2 = \beta_3 := \beta = N + \alpha - \frac{N - \alpha + m}{p + q - m + 1}.$$

Starting with (5.7) and going through the same estimates as above without removing the term $\int_{\mathbb{R}^N} u_0(x)\varphi(x, 0)dx$, the inequality (5.34) changes to

$$\int_0^{2R^\gamma} J(t)^2 dt \leq c_1 R^\beta - c_2 \int_{B_R} u_0(x)dx,$$

where we have used $\psi \equiv 1$ in B_R. Thanks to (5.44), we arrive to

$$\int_0^{2R^\gamma} J(t)^2 dt \leq c\left(R^\beta - R^\nu\right).$$

Since condition (5.45) forces $\beta < \nu$, the above estimate yields a contradiction as $R \to \infty$ in both cases $\nu > 0$ and $\nu = 0$. $\square$

5.3 The Inequality $\frac{\partial u}{\partial t} + \mathscr{L}u \geq (K * u^p)u^q$ in $\mathbb{R}^N \times (0, \infty)$

In this section, we discuss the problem

$$\begin{cases} \frac{\partial u}{\partial t} + \mathscr{L}u \geq (K * u^p)u^q & \text{in } \mathbb{R}^N \times (0, \infty), \\ u(x, 0) = u_0(x) \geq 0 & \text{in } \mathbb{R}^N, N \geq 1. \end{cases} \tag{5.47}$$

Similar to the approach in the previous section, *nonnegative weak solutions* of (5.47) are nonnegative functions $u(x, t)$, belonging to the class $\mathscr{S}$ given by those $u \in W_{\text{loc}}^{1,m}(\mathbb{R}^N \times (0, \infty))$ which fulfil:

(i) $\mathscr{A}(x, u, \nabla u) \in [L_{\text{loc}}^{m'}(\mathbb{R}^N \times (0, \infty))]^N$,

(ii) $(K * u^p)u^q \in L_{\text{loc}}^1(\mathbb{R}^N \times (0, \infty))$,

and such that for any nonnegative test function $\varphi \in C_c^1(\mathbb{R}^N \times \mathbb{R})$, we have

$$\int_0^\infty \int_{\mathbb{R}^N} (K * u^p)u^q \varphi \, dx \, dt \leq \int_0^\infty \int_{\mathbb{R}^N} \frac{\partial u}{\partial t} \varphi dx dt - \int_0^\infty \int_{\mathbb{R}^N} \mathscr{A}(x, u, \nabla u) \cdot \nabla\varphi dx dt.$$

$$\tag{5.48}$$

In this setting, we derive our a priori estimates employing similar arguments to those we used in the proof of Theorem 5.1. Unlike the approach for (5.5) where the a priori estimates are obtained by choosing two classes of test functions in (5.6), for the counterpart problem (5.47), we can only use a single class of test functions. We shall overcome this fact by imposing a higher locally integrability on the solution. The precise solution space will be defined in what follows.

Assume

$$0 < \alpha < m/2 \quad \text{and} \quad p + q > 2 \max\{1, m - 1\}. \tag{5.49}$$

Let $d > 0$ be such that

$$2 < F(d) := \max\left\{\frac{p+q+d}{d+1}, \frac{p+q+d}{d+m-1}\right\} < \frac{2N+m}{N+\alpha}. \tag{5.50}$$

Such a value $d > 0$ always exists since F is decreasing as a function of d and

$$F(0) > 2 \quad \text{and} \quad \lim_{d \to \infty} F(d) = 1 < \frac{2N+m}{N+\alpha}.$$

Our main result concerning the problem (5.47) is the following.

Theorem 5.9 *If (5.49) and (5.50) hold, then problem (5.47) does not have nonnegative nontrivial solutions u in the class*

$$\mathscr{S} \cap \left\{(K * u^p)u^{q+d} \in L^1_{loc}(\mathbb{R}^N \times [0, \infty))\right\}.$$

In particular, if (5.49) holds, then (5.47) has no nonnegative nontrivial solutions $u \in \mathscr{S} \cap C(\mathbb{R}^N \times (0, \infty))$.

Proof Suppose by contradiction that (5.47) admits a nonnegative nontrivial solution $u \in \mathscr{S}$ such that

$$(K * u^p)u^{q+d} \in L^1_{loc}(\mathbb{R}^N \times [0, \infty)). \tag{5.51}$$

In the same way as in Lemma 5.4 (we only have to replace the exponent q by $q+d$), we deduce

$$u^{(p+q+d)/2} \in L^1_{loc}(\mathbb{R}^N \times [0, \infty)).$$

From (5.50) we have

$$\frac{p+q+d}{2} > \max\{d+1, d+m-1\}.$$

Thus, by Hölder's inequality, we deduce

$$u^{d+m-1}, \; u^{d+1} \in L^1_{loc}(\mathbb{R}^N \times [0, \infty)).$$

This will ensure that all integrals in this section are finite.   □

We start with the following result which is a counterpart of Lemma 5.5.

Lemma 5.10 *Let $u \in \mathcal{S}$ be a nonnegative solution of (5.47) satisfying (5.51) and let ψ be defined by (5.15). Then,*

$$\int_0^\infty \int_{\mathbb{R}^N} (K * u^p)u^{q+d}\psi^k \, dx \, dt + \int_0^\infty \int_{\mathbb{R}^N} u^{d-1}\psi^k |\mathscr{A}(x, u, \nabla u)|^{m'} \, dx \, dt$$

$$\leq c_1 \int_0^\infty \int_{\mathbb{R}^N} u^{d+1}\psi^{k-1}\left|\frac{\partial \psi}{\partial t}\right| dx dt + c_2 \int_0^\infty \int_{\mathbb{R}^N} u^{d+m-1}\psi^{k-m}|\nabla\psi|^m dx \, dt,$$

(5.52)

for some constants $c_1, c_2 > 0$ and $d > 0$ satisfies (5.50).

The proof follows line by line that of Lemma 5.5 in which we replace d by $-d$.

Similar to the estimate (5.36), we find

$$\int_{\mathbb{R}^N} (K * u^p)u^{q+d}\psi^k \, dx \, dt \geq R^{-\alpha}L(t)^2,$$ (5.53)

where

$$L(t) = \int_{\mathbb{R}^N} u^\tau(x, t)\psi^k(x, t)dx,$$ (5.54)

where $\tau = (p + q + d)/2 > 1$ and ψ is defined by (5.15). In the same way as we deduced (5.28), by Hölder's inequality , we find

$$\int_0^\infty \int_{\mathbb{R}^N} u^{d+1}\psi^{k-1}\left|\frac{\partial \psi}{\partial t}\right| dx dt \leq \left(\int_0^\infty \int_{\mathbb{R}^N} u^\tau\psi^k\right)^{1/\sigma}\left(\int_0^\infty \int_{\mathbb{R}^N} \psi^{k-\sigma'}\left|\frac{\partial \psi}{\partial t}\right|^{\sigma'}\right)^{1/\sigma'}$$

$$= \left(\int_0^{2R^\gamma} L(t)dt\right)^{1/\sigma}\left(\int_0^\infty \int_{\mathbb{R}^N} \psi^{k-\sigma'}\left|\frac{\partial \psi}{\partial t}\right|^{\sigma'}\right)^{1/\sigma'}$$

$$\leq cR^{\gamma/(2\sigma)}\left(\int_0^{2R^\gamma} L(t)^2 dt\right)^{1/(2\sigma)}\left(\int_0^\infty \int_{\mathbb{R}^N} \psi^{k-\sigma'}\left|\frac{\partial \psi}{\partial t}\right|^{\sigma'}\right)^{1/\sigma'}$$

$$\leq CR^{\frac{2N+\gamma}{2\sigma'} - \frac{\gamma}{2}}\left(\int_0^{2R^\gamma} L(t)^2 dt\right)^{1/(2\sigma)},$$

(5.55)

where $c > 0$ is a constant and

$$\sigma = \frac{\tau}{d+1}, \qquad \sigma' = \frac{\tau}{\tau-d-1}.$$

Note that since $F(d) > 2$, we have $\sigma > 1$.

By Hölder's inequality, similarly to the estimate (5.31), we have

$$\int_0^\infty \int_{\mathbb{R}^N} u^{d+m-1} \psi^{k-m} |\nabla\psi|^m \le \left(\int_0^\infty \int_{\mathbb{R}^N} u^\tau \psi^k\right)^{1/\theta} \left(\int_0^\infty \int_{\mathbb{R}^N} \psi^{k-m\theta'} |\nabla\psi|^{m\theta'}\right)^{1/\theta'}$$

$$= \left(\int_0^{2R^\gamma} L(t) dt\right)^{1/\theta} \left(\int_0^\infty \int_{\mathbb{R}^N} \psi^{k-m\theta'} |\nabla\psi|^{m\theta'}\right)^{1/\theta'}$$

$$\le cR^{\gamma/(2\theta)} \left(\int_0^{2R^\gamma} L(t)^2 dt\right)^{1/(2\theta)} \left(\int_0^\infty \int_{\mathbb{R}^N} \psi^{k-m\theta'} |\nabla\psi|^{m\theta'}\right)^{1/\theta'}$$

$$\le C\left(\int_0^{2R^\gamma} L(t)^2 dt\right)^{1/(2\theta)} R^{\frac{2N+\gamma}{2\theta'}-m+\frac{\gamma}{2}},$$

$$(5.56)$$

where

$$\theta = \frac{\tau}{d+m-1}, \qquad \theta' = \frac{\tau}{\tau-d-m+1}. \qquad (5.57)$$

Note that from $F(d) > 2$, we derive $\eta > 1$. Take now $\gamma = m$.

Next, we use (5.53), (5.55) and (5.56) in the estimate (5.52) of Lemma 5.10 to deduce

$$\int_0^{2R^\gamma} L(t)^2 dt \le c\left(\int_0^{2R^\gamma} L(t)^2 dt\right)^{\frac{1}{2\sigma}} R^{\beta_1} + c\left(\int_0^{2R^\gamma} L(t)^2 dt\right)^{\frac{1}{2\theta}} R^{\beta_2}, \qquad (5.58)$$

where, by (5.50) and $\gamma = m$, we have

$$\beta_1 = \frac{2N+m}{2\sigma'} - \frac{m-2\alpha}{2} = \frac{(N+\alpha)(p+q+d)-(2N+m)(d+1)}{p+q+d} < 0,$$

$$\beta_2 = \frac{2N+m}{2\theta'} - \frac{m-2\alpha}{2} = \frac{(N+\alpha)(p+q+d)-(2N+m)(d+m-1)}{p+q+d} < 0.$$

We further apply Young's inequality in the right-hand side of (5.58) and obtain

$$\int_0^{2R^\gamma} L(t)^2 dt \le \frac{1}{4} \int_0^{2R^\gamma} L(t)^2 dt + CR^{\beta_1(2\sigma)'} + \frac{1}{4} \int_0^{2R^\gamma} L(t)^2 dt + CR^{\beta_2(2\theta)'},$$

that is,

$$\int_0^{2R^\gamma} L(t)^2 dt \le 2C\left(R^{\beta_1(2\sigma)'} + R^{\beta_2(2\theta)'}\right) \longrightarrow 0 \qquad \text{as } R \to \infty.$$

Proceeding as in the proof of Theorem 5.1, this yields $L(t) \equiv 0$ in $[0, \infty)$, which gives $u \equiv 0$, contradiction.

5.4 Conclusions and Further Remarks

The first nonexistence result for nonnegative and nontrivial solutions of semilinear parabolic problems is due to Fujita [Fuj66], where the following Cauchy problem is investigated:

$$\begin{cases} \dfrac{\partial u}{\partial t} - \Delta u = u^q & \text{in } \mathbb{R}^N \times (0, \infty), \\ u(x, 0) = u_0(x) & \text{in } \mathbb{R}^N. \end{cases} \tag{5.59}$$

Fujita obtained the critical exponent $q_F = 1 + 2/N$ on the existence versus nonexistence of nonnegative nontrivial solutions. Precisely, nonexistence of solutions (i.e. blow up) holds when $1 < q < 1 + 2/N$ and $u_0 \geq 0$ is bounded, while blow up can occur when $q > 1 + 2/N$ depending on the size of u_0. Since then, there have been a number of extensions of Fujita results in many directions. In particular, the result obtained by Fujita was completed, relatively to the critical case, in [Hay73] for $N = 1, 2$ and in [KST77] for $N \geq 3$.

When one moves from the equality to the inequality case, in contrast to elliptic problems, the critical exponents for parabolic equations and inequalities coincide, at least in the semilinear case. This observation is due to Mitidieri and Pohozaev in [MP01] where a powerful tool on a priori estimates is devised.

The Fujita exponent for the m-Laplacian parabolic case

$$\begin{cases} \dfrac{\partial u}{\partial t} - \Delta_m u \geq u^q & \text{in } \mathbb{R}^N \times (0, \infty), \\ u(x, 0) = u_0(x) \geq 0 & \text{in } \mathbb{R}^N, \end{cases} \tag{5.60}$$

is obtained in [MP01] where it is proved that (5.60) has no nonnegative nontrivial solutions for $\max\{1, m - 1\} < q \leq m - 1 + m/N$. This condition forces $m > 2N/(N+1)$.

Nonlocal models describe many natural phenomena, such as the non-Newton flux in the mechanics of fluid, population of biological species and filtration. Concerning nonlocal source problems, Galaktionov and Levine [GL98] investigated positive solutions of a Cauchy problem for the following semilinear parabolic equation with weighted nonlocal sources:

$$\begin{cases} \dfrac{\partial u}{\partial t} = \Delta u^m + \left(\displaystyle\int_{\mathbb{R}^N} K(x) u^p(x, t) dx \right)^{(r-1)/p} u^q \ \ \text{in } \mathbb{R}^N \times (0, \infty), \\ u(x, 0) = u_0(x) \geq 0 \text{ in } \mathbb{R}^N, \end{cases}$$

$$(5.61)$$

where $p, q, r \geq 1$, $m > 1$ and K are functions not necessarily in $L^1(\mathbb{R}^N)$. Other semilinear parabolic problems with nonlocal terms are discussed in [QS07].

This chapter stems from the research work [FG22] and presents a first attempt in solving quasilinear parabolic problems of type (5.5) and (5.47) which feature nonlocal convolution terms. The key point in our approach is to obtain suitable integral estimates for the quantity

$$\int_{\mathbb{R}^N} u^\sigma(x, t) \psi^k(x, t) dx$$

where ψ is a specific test function, $\sigma \geq (p+q)/2$ and $k > 0$. Higher-order evolution equations are discussed in the next chapter.

Chapter 6
Higher-Order Evolution Inequalities with Convolution Terms

6.1 Introduction

Let $N, k, m \geq 1$ be positive integers. In this chapter, we are concerned with the problem

$$
\begin{cases}
\dfrac{\partial^k u}{\partial t^k} + (-\Delta)^m u \geq (K * |u|^p)|u|^q & \text{in } \mathbb{R}^N \times (0, \infty), \\[2mm]
\dfrac{\partial^i u}{\partial t^i}(x, 0) = u_i(x) & \text{in } \mathbb{R}^N, \ 0 \leq i \leq k - 1,
\end{cases}
\tag{6.1}
$$

where $N \geq 1$, $p, q > 0$ and $u_i \in L^1_{\text{loc}}(\mathbb{R}^N)$ for $0 \leq i \leq k - 1$. We assume that $K \in C(\mathbb{R}_+; \mathbb{R}_+)$ satisfies

$$
K(|x|) \in L^1_{\text{loc}}(\mathbb{R}^N)
$$

and

(A) there exists $R_0 > 1$ such that $\inf_{r \in (0,R)} K(r) = K(R)$ for all $R > R_0$.

In particular, condition (A) above implies that K is decreasing on the interval (R_0, ∞). Typical examples of potentials K satisfying the above conditions are the constant functions as well as

$$
K(r) = r^{-\alpha}, \ \alpha \in (0, N) \quad \text{or} \quad K(r) = r^{-\alpha} \log^\beta(1+r), \ \alpha \in (0, N), \beta \in \mathbb{R}, \beta > \alpha - N.
$$

By $K * |u|^p$ we denote the standard convolution operator with respect to the space variable, that is,

© The Author(s), under exclusive license to Springer Nature Switzerland AG 2022
M. Ghergu, *Partial Differential Inequalities with Nonlinear Convolution Terms*, SpringerBriefs in Mathematics, https://doi.org/10.1007/978-3-031-21856-9_6

$$(K * |u|^p)(x,t) = \int_{\mathbb{R}^N} K(|x - y|)|u(y,t)|^p dy \quad \text{for all } (x,t) \in \mathbb{R}^N \times (0, \infty).$$

We are interested in *weak solutions* of (6.1), that is, functions $u \in L^p_{\text{loc}}(\mathbb{R}^N \times (0, \infty))$ such that

(i) $(K * |u|^p)|u|^q \in L^1_{\text{loc}}(\mathbb{R}^N \times (0, \infty))$;

(ii) for any nonnegative test function $\varphi \in C^\infty_c(\mathbb{R}^N \times (0, \infty))$, we have

$$\sum_{i=1}^k (-1)^i \int_{\mathbb{R}^N} u_{k-i}(x) \frac{\partial^{i-1} \varphi}{\partial t^{i-1}}(x,0)dx + \int_{\mathbb{R}^N \times (0,\infty)} u\left[(-1)^k \frac{\partial^k \varphi}{\partial t^k} + (-\Delta)^m \varphi\right] dx dt$$

$$\geq \int_{\mathbb{R}^N \times (0,\infty)} (K * |u|^p)|u|^q \varphi \, dx dt. \tag{6.2}$$

Using a standard integration by parts, it is easily seen that any classical solution of of (6.1) is also a weak solution. Let us point out that condition $u \in L^p_{\text{loc}}(\mathbb{R}^N \times (0, \infty))$ follows from the fact that $K * |u|^p$ is finite almost everywhere. Indeed for $R > R_0$ and $x, y \in B_R$, we have $|x - y| \leq 2R$, so that by the definition of K and its monotonicity, we deduce

$$\infty > (K * |u|^p)(x,t) \geq K(2R) \int_{B_R} |u(y,t)|^p dy.$$

6.2 The Main Result

Our main result concerning the inequality (6.1) reads as follows.

Theorem 6.1 *Assume $N, m, k \geq 1$ and $p, q > 0$.*

(i) *If $k \geq 1$ is a even integer and $q \geq 1$, then (6.1) admits positive solutions $u \in C^\infty(\mathbb{R}^N \times (0, \infty))$ which verify*

$$u_{k-1} = \frac{\partial^{k-1} u}{\partial t^{k-1}}(\cdot, 0) < 0 \quad in \ \mathbb{R}^N.$$

(ii) *If $p + q > 2$ and*

$$\limsup_{R \to \infty} K(R) R^{\frac{2N+2m/k}{p+q} - N + 2m(1-1/k)} > 0, \tag{6.3}$$

then (6.1) *has no nontrivial solutions such that*

$$u_{k-1} \geq 0 \quad or \quad u_{k-1} \in L^1(\mathbb{R}^N) \quad and \quad \int_{\mathbb{R}^N} u_{k-1}(x)dx > 0. \tag{6.4}$$

Let us note that condition (6.3) and the fact that K is decreasing in a neighbourhood of infinity imply that

$$\frac{2N + 2m/k}{p+q} \geq N - 2m\left(1 - \frac{1}{k}\right).$$

Also, under extra assumptions on u_{k-1}, we can handle the case $\int_{\mathbb{R}^N} u_{k-1}(x) = 0$ in (6.4) for which the same conclusion as in Theorem 6.1(ii) holds (see Remark 6.4).

Proof

(i) Let $\gamma > 0$ be such that $p\gamma > N$ and define

$$u(x,t) = e^{-Mt}v(x)^{-\gamma/2} \quad with \quad v(x) = 1 + |x|^2, \tag{6.5}$$

where $M > 1$ will be specified later. Since $K > 0$ is continuous in $\mathbb{R}^+$ and decreasing in a neighbourhood of infinity (by condition (A)), it follows that

$$\sup_{[1,\infty)} K = \max_{[1,\infty)} K < \infty.$$

Furthermore, we have

$$
\begin{aligned}
\left(K * v^{-p\gamma/2}\right)(x) &= \int_{\mathbb{R}^N} K(|z|)(1 + |z - x|^2)^{-p\gamma/2} dz \\
&\leq \int_{B_1} K(|z|)dz + \left(\max_{[1,\infty)} K\right) \int_{\mathbb{R}^N \setminus B_1} (1 + |z - x|^2)^{-p\gamma/2} dz \\
&\leq \int_{B_1} K(|z|)dz + \left(\max_{[1,\infty)} K\right) \int_{\mathbb{R}^N} (1 + |z - x|^2)^{-p\gamma/2} dz \\
&= \int_{B_1} K(|z|)dz + \left(\max_{[1,\infty)} K\right) \int_{\mathbb{R}^N} (1 + |y|^2)^{-p\gamma/2} dy \\
&\leq C(K, p, \gamma), \tag{6.6}
\end{aligned}
$$

since $p\gamma > N$.

Further, a direct calculation shows that $-\Delta\left(v^{-\gamma/2}\right) = c_1 v^{-\gamma/2-1} + c_2 v^{-\gamma/2-2}$ in $\mathbb{R}^N$, where c_1, c_2 are real coefficients depending on N and γ. Hence, an induction argument yields

$$(-\Delta)^m \left(v^{-\gamma/2}\right) = v^{-2m-\gamma/2} \sum_{j=0}^{m} c_j v^j \quad \text{in } \mathbb{R}^N,$$

where $c_j = c_j(\gamma, N, m) \in \mathbb{R}$. Thus, the function u given by (6.5) satisfies

$$\frac{\partial^k u}{\partial t^k} + (-\Delta)^m u = e^{-Mt} \left((-1)^k M^k v^{2m} + \sum_{j=0}^{m} c_j v^j\right) v^{-2m-\gamma/2} \quad \text{in } \mathbb{R}^N \times (0, \infty),$$

$$(6.7)$$

where $c_j \in \mathbb{R}$ are independent of M. Using the fact that k is an even integer, by taking $M > 1$ large, we may ensure that

$$M^k v^{2m} + \sum_{j=0}^{m} c_j v^j \geq C(K, p, \gamma) v^{2m} \quad \text{in } \mathbb{R}^N, \tag{6.8}$$

where $C(K, p, \gamma) > 0$ is the constant from (6.6). Now, combining (6.6)–(6.8) and using that $u^p e^{Mpt} = v^{-p\gamma/2}$, we deduce

$$\frac{\partial^k u}{\partial t^k} - \Delta^m u \geq C(K, p, \gamma) e^{-Mt} v^{-\gamma/2}$$

$$\geq C(K, p, \gamma) e^{-(p+q)Mt} v^{-\gamma/2}$$

$$\geq (K * u^p) e^{-qMt} v^{-\gamma/2}$$

$$\geq (K * u^p) \left(e^{-Mt} v^{-\gamma/2}\right)^q$$

$$\geq (K * u^p) u^q \quad \text{in } \mathbb{R}^N \times (0, \infty),$$

which concludes the proof of part (i). Since k is even, one has

$$u_{k-1}(x) = \frac{\partial^{k-1} u}{\partial t^{k-1}}(x, 0) = (-1)^{k-1} M^k v(x) < 0.$$

(ii) Assume that $p + q > 2$, (6.3) and (6.4) hold and that (6.1) admits a weak solution u. We shall first construct a suitable test function φ for (6.2). To do so, take a standard cutoff function $\varrho \in C_c^\infty(\mathbb{R})$ such that:

- $\varrho = 1$ in $(0, 1)$, $\varrho = 0$ in $(2, \infty)$;
- $0 \leq \varrho \leq 1$, $\text{supp}\varrho \subseteq [0, 2]$.

$\square$

Now take $R > 0$, $\gamma > 0$ to be precised later, $\kappa \geq 1$ sufficiently large and consider the function

$$\varphi(x) = \varrho^\kappa\left(\frac{|x|}{R}\right)\varrho^\kappa\left(\frac{t}{R^\gamma}\right) \quad \text{in} \quad \mathbb{R}^N \times (0, \infty). \tag{6.9}$$

Clearly

$$\operatorname{supp}\varphi \subset B_{2R} \times [0, 2R^\gamma) \subset \mathbb{R}^N \times (0, \infty), \tag{6.10}$$

and

$$\operatorname{supp}\frac{\partial^k\varphi}{\partial t^k} \subset B_{2R} \times [R^\gamma, 2R^\gamma) \subset \operatorname{supp}\varphi, \qquad \operatorname{supp}\Delta\varphi \subset (B_{2R}\setminus B_R) \times [0, 2R^\gamma) \subset \operatorname{supp}\varphi. \tag{6.11}$$

We have the following estimates:

Proposition 6.2 *Let φ be defined in (6.9). Then for $\varsigma > 1$ and $\kappa \geq 2m\varsigma$, one has*

$$\int_{\operatorname{supp}\varphi}\frac{1}{\varphi^{\varsigma-1}}\left|\frac{\partial^i\varphi}{\partial t^i}\right|^\varsigma dxdt \leq CR^{N+\gamma-i\gamma\varsigma}, \quad i \geq 1, \tag{6.12}$$

$$\int_{\operatorname{supp}\varphi}\frac{|\Delta^m\varphi|^\varsigma}{\varphi^{\varsigma-1}}dxdt \leq CR^{N+\gamma-2m\varsigma}, \tag{6.13}$$

where C is a positive constant changing from line to line.

Since $\frac{\partial^i\varphi}{\partial t^i}(x, 0) = 0$ in $\mathbb{R}^N$ for all $i \geq 1$, from (6.2) we deduce

$$\int_{\mathbb{R}^N}u_{k-1}(x)\varphi(x, 0)\, dx + \int_{\mathbb{R}^N \times (0,\infty)}(K * |u|^p)|u|^q\varphi\, dxdt$$

$$\leq \int_{\mathbb{R}^N \times (0,\infty)}|u|\left|\frac{\partial^k\varphi}{\partial t^k}\right|dxdt + \int_{\mathbb{R}^N \times (0,\infty)}|u||\Delta^m\varphi|\, dxdt. \tag{6.14}$$

Observe that by (6.4) we have $u_{k-1} \geq 0$ or $u_{k-1} \in L^1(\mathbb{R}^N)$ and $\int_{\mathbb{R}^N}u_{k-1}(x)dx > 0$. In the latter case, from $u_{k-1} \in L^1_{loc}(\mathbb{R}^N)$, we deduce, using (6.9),

$$\lim_{R\to\infty}\int_{\mathbb{R}^N}u_{k-1}(x)\varphi(x, 0)dx = \lim_{R\to\infty}\int_{\mathbb{R}^N}u_{k-1}(x)\varrho^\kappa\left(\frac{|x|}{R}\right)dx = \int_{\mathbb{R}^N}u_{k-1}(x)dx. \tag{6.15}$$

Thus, from (6.4) we deduce that for large $R > 0$ we have

$$\int_{\mathbb{R}^N}u_{k-1}(x)\varphi(x, 0)dx > 0, \tag{6.16}$$

case in which (6.14) yields

$$\int_{\mathbb{R}^N \times (0,\infty)} (K * |u|^p)|u|^q \varphi \, dx dt \leq \int_{\mathbb{R}^N \times (0,\infty)} |u| \left| \frac{\partial^k \varphi}{\partial t^k} \right| dx dt + \int_{\mathbb{R}^N \times (0,\infty)} |u| |\Delta^m \varphi| \, dx dt,$$

$$(6.17)$$

provided $R > 0$ in the definition of φ (see (6.9)) is large enough.

An important tool of our approach is the following result.

Lemma 6.3 *For almost all $t \geq 0$, we have $u(\cdot, t) \in L_{loc}^{\frac{p+q}{2}}(\mathbb{R}^N)$ and*

$$\int_{\mathbb{R}^N} (K * |u|^p)|u(x,t)|^q \varphi(x,t) dx \geq K(4R) J^2(t) \quad \text{for all } R \geq R_0, \qquad (6.18)$$

where

$$J(t) = \int_{\mathbb{R}^N} |u(x,t)|^{\frac{p+q}{2}} \varphi(x,t) dx. \qquad (6.19)$$

Proof First note that for $x, y \in B_{2R}$, one has $|x - y| \leq 4R$, so that, thanks to the monotonicity of K,

$$\int_{\mathbb{R}^N} K(|x-y|)|u(y)|^p dy \geq \int_{B_{2R}} K(|x-y|)|u(y)|^p dy \geq K(4R) \int_{B_{2R}} |u(y)|^p dy.$$

Hence

$$\int_{\mathbb{R}^N} (K * |u|^p)|u(x,t)|^q \varphi(x,t) dx \geq K(4R) \int_{\mathbb{R}^N} \int_{\mathbb{R}^N} |u(y,t)|^p \varphi(y,t)|u(x,t)|^q \varphi(x,t) dx dy,$$

$$(6.20)$$

where we have used that $\varphi \leq 1$ and that $\varphi(\cdot, t) \equiv 0$ outside of B_{2R} for all t.

Furthermore, by Hölder's inequality, we have

$$\left(\iint_{\mathbb{R}^N \times \mathbb{R}^N} |u(y,t)|^p \varphi(y,t)|u(x,t)|^q \varphi(x,t) \, dx \, dy \right)^2$$

$$= \left(\iint_{\mathbb{R}^N \times \mathbb{R}^N} |u(y,t)|^p \varphi(y,t)|u(x,t)|^q \varphi(x,t) \, dx \, dy \right)$$

$$\cdot \left(\iint_{\mathbb{R}^N \times \mathbb{R}^N} |u(x,t)|^p \varphi(x,t)|u(y,t)|^q \varphi(y,t) \, dx \, dy \right)$$

$$\geq \left(\iint_{\mathbb{R}^N \times \mathbb{R}^N} |u(x,t)|^{\frac{p+q}{2}} |u(y,t)|^{\frac{p+q}{2}} \varphi(x,t)\varphi(y,t) \, dx \, dy \right)^2$$

$$= \left(\int_{\mathbb{R}^N} |u(x,t)|^{\frac{p+q}{2}} \varphi(x,t) \, dx \right)^4 = J(t)^4,$$

which, by (6.20) and part (i) in the definition of a solution, yields $u(\cdot, t) \in L_{loc}^{\frac{p+q}{2}}(\mathbb{R}^N)$ and (6.18). $\qquad \square$

Inserting (6.18) into (6.17), we find

$$K(4R) \int_0^{2R^\gamma} J^2(t)dt \le \int_{\mathbb{R}^N \times (0,\infty)} |u| \left(\left| \frac{\partial^k \varphi}{\partial t^k} \right| + |\Delta^m \varphi| \right) dxdt. \qquad (6.21)$$

We next estimate the integral term on the right-hand side of (6.21). Using Hölder's inequality, we have

$$\int_{\mathbb{R}^N \times (0,\infty)} |u| \left| \frac{\partial^k \varphi}{\partial t^k} \right| dxdt \le \left(\int_{\mathrm{supp}(\frac{\partial^k \varphi}{\partial t^k})} |u|^{\frac{p+q}{2}} \varphi \, dxdt \right)^{\frac{2}{p+q}}$$

$$\cdot \left(\int_{\mathrm{supp}(\frac{\partial^k \varphi}{\partial t^k})} \varphi^{-\frac{2}{p+q-2}} \left| \frac{\partial^k \varphi}{\partial t^k} \right|^{\frac{p+q}{p+q-2}} \right)^{\frac{p+q-2}{p+q}}$$

$$\le C \left(\int_{\mathrm{supp}(\frac{\partial^k \varphi}{\partial t^k})} |u|^{\frac{p+q}{2}} \varphi \, dxdt \right)^{\frac{2}{p+q}} R^{-k\gamma + (N+\gamma)\frac{p+q-2}{p+q}},$$

$$(6.22)$$

where in the last inequality, we have used (6.12) with $\varsigma = \frac{p+q}{p+q-2}$, $i = k$ and thanks to (6.11). Note that

$$\mathrm{supp} \left(\frac{\partial^k \varphi}{\partial t^k} \right) \subset B_{2R} \times [R^\gamma, 2R^\gamma]. \qquad (6.23)$$

By Hölder's inequality, we find

$$\int_{\mathbb{R}^N \times (0,\infty)} |u| \left| \frac{\partial^k \varphi}{\partial t^k} \right| dxdt \le C R^{-k\gamma + (N+\gamma)\frac{p+q-2}{p+q}} \left(\int_{R^\gamma}^{2R^\gamma} \left(\int_{\mathbb{R}^N} |u|^{\frac{p+q}{2}} \varphi \, dx \right) dt \right)^{\frac{2}{p+q}}$$

$$\le C R^{-k\gamma + (N+\gamma)\frac{p+q-2}{p+q} + \frac{\gamma}{p+q}} \left(\int_{R^\gamma}^{2R^\gamma} \left(\int_{\mathbb{R}^N} |u|^{\frac{p+q}{2}} \varphi \, dx \right)^2 dt \right)^{\frac{1}{p+q}}.$$

$$(6.24)$$

Similarly, by Hölder's inequality, (6.13) with $\varsigma = \frac{p+q}{p+q-2}$ and (6.11), we find

$$\int_{\mathbb{R}^N \times (0,\infty)} |u| |\Delta^m \varphi| dxdt \le \left(\int_{\mathrm{supp}(\Delta^m \varphi)} |u|^{\frac{p+q}{2}} \varphi \, dxdt \right)^{\frac{2}{p+q}}$$

$$\cdot \left(\int_{\mathrm{supp}(\Delta^m \varphi)} \varphi^{-\frac{2}{p+q-2}} |\Delta^m \varphi|^{\frac{p+q}{p+q-2}} \right)^{\frac{p+q-2}{p+q}}$$

$$\le C \left(\int_{\mathrm{supp}(\Delta^m \varphi)} |u|^{\frac{p+q}{2}} \varphi \, dxdt \right)^{\frac{2}{p+q}} R^{-2m + (N+\gamma)\frac{p+q-2}{p+q}}.$$

$$(6.25)$$

Since

$$\text{supp}(\Delta^m \varphi) \subset (B_{2R} \setminus B_R) \times [0, 2R^\gamma], \tag{6.26}$$

a new application of Hölder's inequality in the above estimate yields

$$\int_{\mathbb{R}^N \times (0,\infty)} |u| |\Delta^m \varphi| dx dt \leq C R^{-2m + (N+\gamma)\frac{p+q-2}{p+q}} \left(\int_0^{2R^\gamma} \left(\int_{B_{2R} \setminus B_R} |u|^{\frac{p+q}{2}} \varphi \, dx \right) dt \right)^{\frac{2}{p+q}}$$

$$\leq C R^{-2m + (N+\gamma)\frac{p+q-2}{p+q} + \frac{\gamma}{p+q}} \left(\int_0^{2R^\gamma} \left(\int_{B_{2R} \setminus B_R} |u|^{\frac{p+q}{2}} \varphi \, dx \right)^2 dt \right)^{\frac{1}{p+q}}. \tag{6.27}$$

Comparing the powers of R in (6.24) and (6.27), we are led to choose $\gamma > 0$ so that $k\gamma = 2m$, that is, $\gamma = 2m/k$. Also,

$$-2m + (N+\gamma)\frac{p+q-2}{p+q} + \frac{\gamma}{p+q} = N - 2m\left(1 - \frac{1}{k}\right) - \frac{2N + \frac{2m}{k}}{p+q}.$$

Thus, (6.21) together with (6.24) and (6.27) yields

$$K(4R) \int_0^{2R^\gamma} J^2(t) dt \leq C R^{N - 2m\left(1 - \frac{1}{k}\right) - \frac{2N + 2m/k}{p+q}} \times$$

$$\times \left[\left(\int_{R^\gamma}^{2R^\gamma} \left(\int_{\mathbb{R}^N} |u|^{\frac{p+q}{2}} \varphi \, dx \right)^2 dt \right)^{\frac{1}{p+q}} + \left(\int_0^{2R^\gamma} \left(\int_{B_{2R} \setminus B_R} |u|^{\frac{p+q}{2}} \varphi \, dx \right)^2 dt \right)^{\frac{1}{p+q}} \right]. \tag{6.28}$$

Using (6.19), the above estimate implies

$$K(4R) \int_0^{2R^\gamma} J^2(t) dt \leq C R^{N - 2m\left(1 - \frac{1}{k}\right) - \frac{2N + 2m/k}{p+q}} \left(\int_0^{2R^\gamma} J^2(t) dt \right)^{\frac{1}{p+q}} \tag{6.29}$$

which further yields

$$\left(\int_0^{2R^\gamma} J^2(t) dt \right)^{\frac{p+q-1}{p+q}} \leq C \frac{1}{K(4R) R^{\frac{2N + 2m/k}{p+q} - N + 2m(1-1/k)}}. \tag{6.30}$$

Let $\{R_j\}_j$ be a divergent sequence that achieves the limsup in (6.3), namely,

$$K(4R_j) R_j^{\frac{2N + 2m/k}{p+q} - N + 2m(1-1/k)} \to \ell > 0 \quad \text{as} \quad j \to \infty.$$

Passing to a subsequence, we may assume $R_j > 2R_{j-1}$ for all $j > 1$.

If $\ell = \infty$, we can pass to the limit in (6.30), by replacing R with R_j, to raise $\int_0^\infty J^2(t) dt = 0$ from where $J \equiv 0$ in $\mathbb{R}^+$ and then, by the definition of J,

$$\int_{B_R \times (0,\infty)} |u|^{(p+q)/2} dx dt = 0 \quad \text{for all} \quad R > 0,$$

namely, $u \equiv 0$ in $\mathbb{R}^N \times (0, \infty)$ as required.

If $\ell \in (0, \infty)$, then (6.30) shows that $J \in L^2(0, \infty)$. Using this fact, we infer that

$$\int_{R_j^m}^{2R_j^m} \left(\int_{\mathbb{R}^N} |u(x,t)|^{\frac{p+q}{2}} \varphi(x,t)dx \right)^2 dt \to 0, \quad \int_0^{2R_j^m} \left(\int_{B_{2R_j}\backslash B_{R_j}} |u(x,t)|^{\frac{p+q}{2}} \varphi(x,t)dx \right)^2 dt \to 0,$$
(6.31)

as $j \to \infty$. Indeed, the first limit in (6.31) follows immediately by the fact that $J \in L^2(0, \infty)$. To check the second limit in (6.31), we observe that

$$\infty > \int_0^\infty \left(\int_{\mathbb{R}^N} |u(x,t)|^{\frac{p+q}{2}} dx \right)^2 dt \geq \int_0^\infty \left(\sum_{j\geq 1} \int_{B_{2R_j}\backslash B_{R_j}} |u(x,t)|^{\frac{p+q}{2}} dx \right)^2 dt$$

$$\geq \int_0^\infty \sum_{j\geq 1} \left(\int_{B_{2R_j}\backslash B_{R_j}} |u(x,t)|^{\frac{p+q}{2}} dx \right)^2 dt$$

$$= \sum_{j\geq 1} \int_0^\infty \left(\int_{B_{2R_j}\backslash B_{R_j}} |u(x,t)|^{\frac{p+q}{2}} dx \right)^2 dt.$$

The convergence of the last series in the above estimate implies the second part of (6.31). Using this fact in (6.28), we find

$$\left(\int_0^{2R_j^\gamma} J^2(t)dt \right)^{\frac{p+q-2}{p+q}} \leq C \frac{1}{K(4R_j) R_j^{\frac{2N+2m/k}{p+q}-N+2m(1-1/k)}} o(1) \quad \text{as } j \to \infty.$$

Again, by letting $R_j \to \infty$, we can conclude, also when $\ell \in (0, \infty)$, that $\int_0^\infty J^2(t)dt = 0$, namely, $u \equiv 0$ in $\mathbb{R}^N \times (0, \infty)$.

A similar argument allows us to treat the case $\int_{\mathbb{R}^N} u_{k-1}(x)dx = 0$. In this case, we need to be more precise on the behaviour of u_{k-1}.

Proposition 6.4 *Let $\varrho \in C_c^\infty(\mathbb{R})$ satisfy supp $\varrho \subseteq [0, 2]$, $0 \leq \varrho \leq 1$ and $\varrho = 1$ in $(0, 1)$.*

Assume that for some $\kappa \geq 2m$, we have

$$\int_{\mathbb{R}^N} u_{k-1}(x)\varrho^\kappa \left(\frac{x}{R} \right) dx = O(R^{-\beta}) \quad \text{as } R \to \infty, \quad \text{for some } \beta > 0. \quad (6.32)$$

If

$$\limsup_{R\to\infty} K(R) R^{\min\{\beta, \frac{2N+2m/k}{p+q}-N+2m(1-1/k)\}} > 0, \quad (6.33)$$

then (6.1) has no nontrivial solutions.

Proof With the help of the above ϱ, we construct the test function φ as defined in (6.9). Thus, (6.16) is no more in force and (6.14) changes to

$$\int_{\mathbb{R}^N \times (0,\infty)} (K * |u|^p)|u|^q \varphi \, dxdt \leq CR^{-\beta} + \int_{\mathbb{R}^N \times (0,\infty)} |u| \left|\frac{\partial^k \varphi}{\partial t^k}\right| dxdt + \int_{\mathbb{R}^N \times (0,\infty)} |u| |\Delta^m \varphi| \, dxdt.$$

Consequently using (6.24) and (6.27), with $\gamma = 2k/m$, and (6.18), then the above inequality gives

$$K(4R) \int_0^{2R^\gamma} J^2(t)dt \leq CR^{-\beta} + CR^{N-2m\left(1-\frac{1}{k}\right) - \frac{2N+2m/k}{p+q}} \left(\int_0^{2R^\gamma} J^2(t)dt\right)^{\frac{1}{p+q}}. \tag{6.34}$$

Let $\sigma = \frac{2N+2m/k}{p+q} - N + 2m(1 - 1/k)$. By Young's inequality, we have

$$CR^{N-2m\left(1-\frac{1}{k}\right) - \frac{2N+2m/k}{p+q}} \left(\int_0^{2R^\gamma} J^2(t)dt\right)^{\frac{1}{p+q}} = CR^{-\sigma} \left(\int_0^{2R^\gamma} J^2(t)dt\right)^{\frac{1}{p+q}}$$

$$= CR^{-\sigma} K(4R)^{-\frac{1}{p+q}} \left(K(4R) \int_0^{2R^\gamma} J^2(t)dt\right)^{\frac{1}{p+q}}$$

$$\leq \frac{K(4R)}{2} \int_0^{2R^\gamma} J^2(t)dt + CR^{-\frac{\sigma(p+q)}{p+q-1}} K(4R)^{-\frac{1}{p+q-1}}.$$

Using this last inequality into (6.34), we find

$$\frac{K(4R)}{2} \int_0^{2R^\gamma} J^2(t)dt \leq CR^{-\beta} + CR^{-\frac{\sigma(p+q)}{p+q-1}} K(4R)^{-\frac{1}{p+q-1}},$$

that is,

$$\int_0^{2R^\gamma} J^2(t)dt \leq \frac{C}{R^\beta K(4R)} + \frac{C}{\left(R^\sigma K(4R)\right)^{\frac{p+q}{p+q-1}}}.$$

Now, in virtue of (6.33), we may let $R \to \infty$ in the above estimate to deduce $J = 0$ and then $u \equiv 0$. $\qquad\square$

6.3 Two Consequences of the Main Result

Theorem 6.1 shows a sharp contrast in the existence/nonexistence diagram according to whether $\frac{\partial^{k-1}u}{\partial t^{k-1}}(x, 0)$ has constant sign (positive or negative) on $\mathbb{R}^N$. To better illustrate this fact, let us discuss the case of pure powers in the potential $K(r) = r^{-\alpha}$, $\alpha \in (0, N)$ and $k = 1, 2$.

Let us first consider the parabolic problem

$$\begin{cases} \dfrac{\partial u}{\partial t} + (-\Delta)^m u \geq (|x|^{-\alpha} * |u|^p)|u|^q & \text{in } \mathbb{R}^N \times (0, \infty), \\ u(x, 0) = u_0(x) \text{ in } \mathbb{R}^N. \end{cases} \tag{6.35}$$

Note that since $k = 1$ in (6.1), condition (6.4) is satisfied for all nonnegative solutions of (5.5).

Corollary 6.5 *Let $N, m \geq 1$, $p, q > 0$ and $\alpha \in (0, N)$.*

(i) *If $0 < \alpha < m$ and*

$$2 < p + q \leq \frac{2N + 2m}{N + \alpha},$$

then (6.1) has no nontrivial nonnegative solutions;
(ii) *If $N > 2m$ and*

$$\min\{p, q\} > \max\left\{1, \frac{N - \alpha}{N - 2m}\right\} \text{ and } p + q > \frac{2N - \alpha}{N - 2m},$$

then (5.5) has positive solutions.

Part (i) in the above result follows from Theorem 6.1, while part (ii) follows from Theorem 4.10 where the stationary case of (5.5) is discussed. Let us note that for $m = 1$, the nonexistence of a nonnegative solution in Corollary 6.5(i) was already observed in Corollary 5.3.

We next take $K(r) = r^{-\alpha}$, $k = 2$ and $m = 1$ in Theorem 6.1. We thus consider

$$\begin{cases} \dfrac{\partial^2 u}{\partial t^2} - \Delta u \geq (|x|^\alpha * |u|^p)|u|^q & \text{in } \mathbb{R}^N \times (0, \infty) := \mathbb{R}^N \times \mathbb{R}_+, \\ u(x, 0) = u_0(x) \text{ in } \mathbb{R}^N, \\ \dfrac{\partial u}{\partial t}(x, 0) = u_1(x) \text{ in } \mathbb{R}^N. \end{cases} \tag{6.36}$$

Our result on problem (6.36) is stated below.

Corollary 6.6 *Let $N, m \geq 1$, $p, q > 0$ and $\alpha \in (0, N)$.*

(i) *If $q \geq 1$, then (6.1) admits positive solutions $u \in C^2(\mathbb{R}_+^{N+1})$ which verify*

$$\frac{\partial u}{\partial t}(\cdot, 0) < 0 \quad \text{in } \mathbb{R}^N.$$

(ii) *If $0 < \alpha < \min\{N, 3/2\}$ and*

$$2 < p + q \leq \frac{2N + 1}{N + \alpha - 1},$$

then (6.36) *has no nontrivial solutions satisfying*

$$\frac{\partial u}{\partial t}(x, \cdot) \geq 0 \ in \ \mathbb{R}^N \quad or \quad \int_{\mathbb{R}^N} \frac{\partial u}{\partial t}(x, \cdot) dx > 0. \tag{6.37}$$

(iii) *If* $N > 2$ *and*

$$\min\{p, q\} > \frac{N - \alpha}{N - 2} \quad and \quad p + q > \frac{2N - \alpha}{N - 2}, \tag{6.38}$$

then (6.36) *admits positive solutions which verify* $\dfrac{\partial u}{\partial t}(x, \cdot) > 0 \ in \ \mathbb{R}^N.$

The diagram of existence/nonexistence of a weak solution to (6.36) satisfying (6.37) in the pq-plane is given below. The light shaded region represents the region for which (6.36) admits solutions satisfying (6.37), while the dark shaded region describes the pairs (p, q) for which no such solutions exist (Fig. 6.1).

Corollary 6.6 leaves open the isue of existence and nonexistence in the white regions of the $(p > 0, q > 0)$ quadrant.

Theorem 6.1 also applies to the case where $K \equiv 1$ for which (6.1) reads

$$\begin{cases} \dfrac{\partial^k u}{\partial t^k} - \Delta^m u \geq \left(\displaystyle\int_{\mathbb{R}^N} |u(y)|^p dy \right) |u|^q & in \ \mathbb{R}^N \times (0, \infty), \\ \dfrac{\partial^i u}{\partial t^i}(x, 0) = u_i(x) & in \ \mathbb{R}^N, \ 0 \leq i \leq k - 1. \end{cases} \tag{6.39}$$

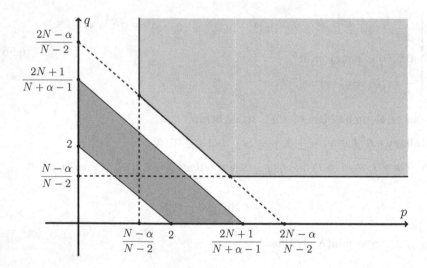

Fig. 6.1 The existence region (light shaded) and the nonexistence region (dark shaded) for weak solutions of (6.36)

A similar conclusion to Corollary 6.5 and Corollary 6.6 (in which we let $\alpha = 0$) holds for (6.39).

Proof Part (i) and (ii) in Corollary 6.6 follow directly from Theorem 6.1.
  (iii) Let p, q, α satisfy (6.38) which we may write as

$$(N - 2)\min\{p, q\} > N - \alpha \quad \text{and} \quad (N - 2)(p + q - 1) > N + 2 - \alpha.$$

Thus, we may choose $\beta \in (0, N - 2)$ such that

$$\beta \min\{p, q\} > N - \alpha, \quad \beta p \neq N \quad \text{and} \quad \beta(p + q - 1) > N + 2 - \alpha \quad (6.40)$$

Set $w(x, t) = (1 + t)^{-a} + 1 + |x|^2$ where $a > 0$ will be precised later and $u(x, t) = w^{-\beta/2}(x, y)$. Then

$$-\Delta u = \beta N w^{-\beta/2-1} - \beta(\beta + 2)w^{-\beta/2-2}|x|^2$$

$$= \beta\left[N - (\beta + 2)\frac{|x|^2}{w}\right]w^{-\beta/2-1} \quad (6.41)$$

$$\geq \beta(N - \beta - 2)w^{-(\beta+2)/2} \quad \text{in } \mathbb{R}^N \times (0, \infty).$$

Also,

$$\frac{\partial^2 u}{\partial t^2} = -\frac{a(a + 1)\beta}{2}(1 + t)^{-a-2}w^{-(\beta+2)/2} + \frac{a^2\beta(\beta + 2)}{4}(1 + t)^{-2a-2}w^{-(\beta+2)/2-1}$$

$$= \frac{a\beta}{2}w^{-(\beta+2)/2}(1 + t)^{-a-2}\left[-a - 1 + \frac{a}{2}(\beta + 2)\frac{(1 + t)^a}{w}\right]$$

$$\geq -\frac{a(a + 1)\beta}{2}w^{-(\beta+2)/2} \quad \text{in } \mathbb{R}^N \times (0, \infty).$$

$$(6.42)$$

Combining now (6.41) and (6.42), we have

$$\frac{\partial^2 u}{\partial t^2} - \Delta u \geq \beta\left(N - 2 - \beta - \frac{a(a + 1)}{2}\right)w^{-(\beta+2)/2} \quad \text{in } \mathbb{R}^N \times (0, \infty).$$

Thus, by letting $a > 0$ small enough, the above estimate yields

$$\frac{\partial^2 u}{\partial t^2} - \Delta u \geq Cw^{-(\beta+2)/2} \quad \text{in } \mathbb{R}^N \times (0, \infty), \quad (6.43)$$

for some constant $C > 0$. To proceed further, we need the following result.   □

Lemma 6.7 *Let* $\alpha \in (0, N)$, $\sigma > N - \alpha$ *and* $1 \leq \lambda \leq 2$. *Then, there exists a constant* $C = C(N, \alpha, \sigma) > 0$ *(note that C is independent of* λ) *such that*

$$\int_{\mathbb{R}^N} \frac{dy}{|x-y|^\alpha(\lambda+|y|^2)^{\sigma/2}} \le C \begin{cases} (\lambda+|x|^2)^{(N-\alpha-\sigma)/2} & \text{if } \sigma < N, \\ (\lambda+|x|^2)^{-\alpha/2} & \text{if } \sigma > N, \end{cases} \quad \text{for all } |x| \ge 1.$$

Proof Let $|x| \ge 1$. We split the integral into

$$\int_{\mathbb{R}^N} \frac{dy}{|x-y|^\alpha(\lambda+|y|^2)^{\sigma/2}} = \left\{ \int_{|y|\ge 2|x|} + \int_{\frac{1}{2}|x|\le|y|\le 2|x|} + \int_{|y|\le|x|/2} \right\} \frac{dy}{|x-y|^\alpha(\lambda+|y|^2)^{\sigma/2}}.$$

If $|y| \ge 2|x|$, then $|x-y| \ge |y|-|x| \ge |y|/2$ and we have

$$\int_{|y|\ge 2|x|} \frac{dy}{|x-y|^\alpha(\lambda+|y|^2)^{\sigma/2}} \le C \int_{|y|\ge 2|x|} \frac{dy}{|y|^{\alpha+\sigma}} \le C|x|^{N-\alpha-\sigma}$$

$$\le C\left(\frac{\lambda+|x|}{3}\right)^{N-\alpha-\sigma}$$

$$\le C(\lambda+|x|)^{N-\alpha-\sigma}$$

$$\le C(\lambda+|x|^2)^{(N-\alpha-\sigma)/2},$$

where we have used that

$$|x| \ge \frac{\lambda+|x|}{3} \quad \text{if} \quad |x| \ge 1 \quad \text{and} \quad \lambda \le 2. \tag{6.44}$$

Next we have

$$\int_{\frac{1}{2}|x|\le|y|\le 2|x|} \frac{dy}{|x-y|^\alpha(\lambda+|y|^2)^{\sigma/2}} \le C(\lambda+|x|^2)^{-\sigma/2} \int_{\frac{1}{2}|x|\le|y|\le 2|x|} \frac{dy}{|x-y|^\alpha}$$

$$\le C(\lambda+|x|^2)^{-\sigma/2} \int_{|y-x|\le 3|x|} \frac{dy}{|x-y|^\alpha}$$

$$\le C(\lambda+|x|^2)^{-\sigma/2}|x|^{N-\alpha}$$

$$\le C(\lambda+|x|^2)^{(N-\alpha-\sigma)/2},$$

where we have used that $|y| \le 2|x|$ implies $|y-x| \le 3|x|$.

Finally, if $|y| \le |x|/2$ then $|x-y| \ge |x|-|y| \ge |x|/2$. We have

$$\int_{|y|\le|x|/2} \frac{dy}{|x-y|^\alpha(\lambda+|y|^2)^{\sigma/2}} \le C|x|^{-\alpha} \int_{|y|\le|x|/2} \frac{dy}{(\lambda+|y|^2)^{\sigma/2}}. \tag{6.45}$$

If $\sigma < N$ then

$$\int\limits_{|y|\leq|x|/2} \frac{dy}{(\lambda + |y|^2)^{\sigma/2}} \leq \int\limits_{|y|\leq|x|/2} \frac{dy}{|y|^\sigma} = C|x|^{N-\sigma},$$

so that by (6.44) and $N - \alpha - \sigma < 0$, we find

$$\int\limits_{|y|\leq|x|/2} \frac{dy}{|x - y|^\alpha(\lambda + |y|^2)^{\sigma/2}} \leq C|x|^{N-\alpha-\sigma} \leq C\left(\frac{\lambda + |x|}{3}\right)^{N-\alpha-\sigma}$$

$$\leq C(\lambda + |x|^2)^{(N-\alpha-\sigma)/2}.$$

If $\sigma > N$ then

$$\int\limits_{|y|\leq|x|/2} \frac{dy}{(\lambda + |y|^2)^{\sigma/2}} \leq \int\limits_{\mathbb{R}^N} \frac{dy}{(1 + |y|^2)^{\sigma/2}} < C < \infty,$$

and from (6.45) and (6.44), one has

$$\int\limits_{|y|\leq|x|/2} \frac{dy}{|x - y|^\alpha(\lambda + |y|^2)^{\sigma/2}} \leq C|x|^{-\alpha} \leq C\left(\frac{\lambda + |x|}{3}\right)^{-\alpha} \leq C(\lambda + |x|^2)^{-\alpha/2},$$

which concludes our proof.   □

Let us return to the proof of Corollary 6.6 and observe that

$$\left(|x|^{-\alpha} * u^p\right)u^q \leq (1 + |x|^2)^{-\beta q/2} \int\limits_{\mathbb{R}^N} \frac{dy}{|x - y|^\alpha(1 + |y|^2)^{\beta p/2}},$$

for all $(x, t) \in \mathbb{R}^N \times (0, \infty)$, since $w(x, t) \geq 1 + |x|^2$. Since from (6.40) we have $\beta p > 0$ and $\alpha < N$, the above integral is finite and thus $\left(|x|^{-\alpha} * u^p\right)u^q$ is uniformly bounded from above for $(x, t) \in B_1 \times [0, \infty)$. In addition, since $1 \leq w \leq 3$ in $(x, t) \in B_1 \times [0, \infty)$, then by (6.43) $\frac{\partial^2 u}{\partial t^2} - \Delta u$ and $\left(|x|^{-\alpha} * u^p\right)u^q$ are positive, continuous and uniformly bounded functions from below and from above, respectively, on $(x, t) \in B_1 \times [0, \infty)$. We may thus find $C_1 > 0$ such that

$$\frac{\partial^2 u}{\partial t^2} - \Delta u \geq C_1\left(|x|^{-\alpha} * u^p\right)u^q \quad \text{in } B_1 \times [0, \infty). \tag{6.46}$$

On the other hand, by Lemma 6.7 for $\lambda = 1 + (1 + t)^{-a}$ and $\sigma = \beta p$, where $\sigma > N - \alpha$ by (6.40), we have

$$\left(|x|^{-\alpha} * u^p\right)u^q \leq C \begin{cases} w^{\frac{N-\alpha-\beta(p+q)}{2}} & \text{if } \sigma = \beta p < N, \\ w^{-\frac{\beta q+\alpha}{2}} & \text{if } \sigma = \beta p > N, \end{cases}$$

for all $(x, t) \in (\mathbb{R} \setminus B_1) \times [0, \infty)$. Using the above estimate together with (6.40) and (6.43), we find

$$\frac{\partial^2 u}{\partial t^2} - \Delta u \geq Cw^{-(\beta+2)/2} \geq Cw^{\max\{\frac{N-\alpha-\beta(p+q)}{2}, -\frac{\beta q+\alpha}{2}\}} \geq C_2\left(|x|^{-\alpha} * u^p\right)u^q,$$
(6.47)

for all $(x, t) \in (\mathbb{R} \setminus B_1) \times [0, \infty)$. Letting now $M = \left(\max\{C_1, C_2\}\right)^{1/(p+q)-1}$, it follows from (6.46) and (6.47) that $U = Mu$ is a $C^\infty(\mathbb{R}^N \times (0, \infty))$ solution of (6.36) such that

$$\frac{\partial U}{\partial t}(x, 0) = \frac{\alpha\beta M}{2}\left(2 + |x|^2\right)^{-(\beta+2)/2} > 0.$$

6.4   Conclusions and Further Remarks

Since the early 1980s, many research works have been devoted to the study of the prototype evolution inequalities:

$$\frac{\partial u}{\partial t} - \Delta u \geq |u|^q \quad \text{and} \quad \frac{\partial^2 u}{\partial t^2} - \Delta u \geq |u|^q \quad \text{in } \mathbb{R}^N \times (0, \infty).$$

To the best of our knowledge, the first study of nonexistence of solutions to higher-order hyperbolic inequalities is due to L. Véron and S.I. Pohozaev [PV00] related to

$$\frac{\partial^2 u}{\partial t^2} - \mathscr{L}_m\big(\phi(u)\big) \geq |u|^q \quad \text{in } \mathbb{R}^N \times (0, \infty),$$
(6.48)

where

$$\mathscr{L}_m v = \sum_{|\alpha|=m} D^\alpha(a_\alpha(x, t)v) \quad \text{for some integer } m \geq 1$$

and ϕ is a locally bounded real-valued function such that $|\phi(u)| \leq c|u|^p$, for some $c, p > 0$. It is obtained in [PV00] that if $q > \max\{p, 1\}$ and one of the following conditions hold

$$\text{either} \quad m \geq 2N \quad \text{or} \quad m < 2N \leq \frac{m(q+1)}{q-p},$$
(6.49)

then (6.48) has no solutions in $\mathbb{R}^N \times (0, \infty)$ satisfying

$$\int_{\mathbb{R}^N} \frac{\partial u}{\partial t}(x, 0)dx \geq 0. \tag{6.50}$$

Condition (6.50) is essential in the study of nonexistence of solutions to (6.48) and will also play an important role in the study of (6.1) (see Theorem 6.1). Further, if $m = 2$, it is obtained in [PV00] that if (6.49) fails to hold, then (6.48) has a positive solution. Among the results for $m = 2$, we quote also [Gue03], where a weighted nonlinearity is considered.

Another set of results that motivate our present work are due to G.G. Laptev [Lap02, Lap03] where the following problem is studied:

$$\begin{cases} \dfrac{\partial^k u}{\partial t^k} - \Delta u \geq |x|^{-\sigma} |u|^q, u \geq 0 & \text{in } \Omega \times (0, \infty), \\ \dfrac{\partial^{k-1} u}{\partial t^{k-1}}(x, 0) \geq 0 & \text{in } \Omega. \end{cases} \tag{6.51}$$

In the above, $\Omega \subset \mathbb{R}^N$ is either the exterior of a ball or an unbounded cone-like domain; for other results on hyperbolic inequalities in exterior domains, see [JS21, JSY19]. We observe that solutions of (6.51) are also required to satisfy (6.50). It is obtained in [Lap03] that if $\sigma > -2$, then the above problem has no solutions provided that

$$1 < q < q_k^* := 1 + \frac{2+\sigma}{N - 2 + 2/k}.$$

The above exponent q_k^* coincides with the Fujita-Hayakawa critical exponent if $k = 1$, that is, $q_1^* = 1 + \frac{2+\sigma}{N}$, and with the Kato critical exponent if $k = 2$, that is, $q_2^* = 1 + \frac{2+\sigma}{N-1}$.

The results obtained in this chapter can be naturally extended to the case of systems of type

$$\begin{cases} \dfrac{\partial^k u}{\partial t^k} + (-\Delta)^m u \geq (K * |v|^p)|v|^q & \text{in } \mathbb{R}^N \times (0, \infty), \\ \dfrac{\partial^k v}{\partial t^k} + (-\Delta)^m v \geq (L * |u|^n)|u|^s & \text{in } \mathbb{R}^N \times (0, \infty), \\ \dfrac{\partial^i u}{\partial t^i}(x, 0) = u_i(x) & \text{in } \mathbb{R}^N, 0 \leq i \leq k - 1, \\ \dfrac{\partial^i v}{\partial t^i}(x, 0) = v_i(x) & \text{in } \mathbb{R}^N, 0 \leq i \leq k - 1, \end{cases} \tag{6.52}$$

where $N, m, k \geq 1$, $p, q, n, s > 0$, $u_i, v_i \in L^1_{loc}(\mathbb{R}^N)$, $0 \leq i \leq k - 1$ and K, L satisfy condition (A) for some $R_0 > 1$. Using the same approach as in the proof of

Theorem 6.1 one obtains that if $\min\{p+q, n+s\} > 2$ and either

$$\limsup_{R\to\infty} K(R)L(R)^{\frac{1}{n+s}} R^{\frac{2N+2m/k}{(n+s)(p+q)} + \frac{N+2m}{n+s} - N + 2m\left(1-\frac{1}{k}\right)} > 0, \tag{6.53}$$

or

$$\limsup_{R\to\infty} K(R)^{\frac{1}{p+q}} L(R) R^{\frac{2N+2m/k}{(n+s)(p+q)} + \frac{N+2m}{p+q} - N + 2m\left(1-\frac{1}{k}\right)} > 0, \tag{6.54}$$

then the system (6.52) has no nontrivial solutions that satisfy one of the following conditions:

- $u_{k-1}, v_{k-1} \geq 0$;
- $u_{k-1}, v_{k-1} \in L^1(\mathbb{R}^N)$ and $\displaystyle\int_{\mathbb{R}^N} u_{k-1}(x)dx, \int_{\mathbb{R}^N} v_{k-1}(x)dx > 0$.

Appendix A
Some Properties of Superharmonic Functions

In this section, we state two important results on superharmonic functions. For the sake of completeness, we also provide their proofs.

Lemma A.1 *Let $v \in L^1_{loc}(\mathbb{R}^N)$ be a nonnegative superharmonic function in the sense that*[1]

$$v(x) \geq M_R(x) := \fint_{B_R(x)} v(y)dy \quad \text{for a.a. } x \in \mathbb{R}^N \text{ and } R > 0.$$

Then,

$$\lim_{R \to \infty} M_R(x) = \operatorname{essinf}_{\mathbb{R}^N} v \quad \text{for all } x \in \mathbb{R}^N.$$

Proof For any $x \in \mathbb{R}^N$ denote

$$\ell(x) = \liminf_{R \to \infty} M_R(x) \quad \text{and} \quad L(x) = \limsup_{R \to \infty} M_R(x).$$

Let $x, y \in \mathbb{R}^N$, $\delta := |x - y| > 0$ and $R > 0$. Then $B_R(x) \subset B_{R+\delta}(y)$ so that

$$v(y) \geq M_{R+\delta}(y) \geq \left(\frac{R}{R + \delta} \right)^N M_R(x). \tag{A.1}$$

Letting $R \to \infty$, this yields $L(y) \geq L(x)$ and $\ell(y) \geq \ell(x)$. Interchanging x and y, we also have the converse inequality, so

---

[1] By $\fint_A f(y)dy$, we denote the standard average of the function f over $A \subset \mathbb{R}^N$ (with respect to the Lebesgue measure on $\mathbb{R}^N$).

© The Author(s), under exclusive license to Springer Nature Switzerland AG 2022   123
M. Ghergu, *Partial Differential Inequalities with Nonlinear Convolution Terms*,
SpringerBriefs in Mathematics, https://doi.org/10.1007/978-3-031-21856-9

$$\ell(x) = \ell(y) \quad \text{and} \quad L(x) = L(y) \quad \text{for all } x, y \in \mathbb{R}^N.$$

Since $M_R(x) \geq \text{essinf}_{\mathbb{R}^N} v$, it follows that $\ell(x) \geq \text{essinf}_{\mathbb{R}^N} v$. On the other hand, we deduce from (A.1) that $v(y) \geq \ell(x)$ for all $x, y \in \mathbb{R}^N$ so that $\text{essinf}_{\mathbb{R}^N} v \geq \ell(x)$ and finally $\ell(x) = \text{essinf}_{\mathbb{R}^N} v$ which finishes the proof.   □

Lemma A.2 *Let μ be a positive Radon measure on $\mathbb{R}^N$, $N \geq 3$ and*

$$u(x) = \frac{\Gamma\left(\frac{N-2}{2}\right)}{2^2 \pi^{N/2}} \int_{\mathbb{R}^N} |x - y|^{2-N} d\mu(y) \quad \text{for all } x \in \mathbb{R}^N.$$

Assuming that u is finite a.e., then u is superharmonic, $u \in L^1_{loc}(\mathbb{R}^N)$, u satisfies the ring condition (4.6) and $\text{essinf}_{\mathbb{R}^N} u = 0$.

Proof It is a known fact that u is superharmonic and $u \in L^1_{loc}(\mathbb{R}^N)$; see, for instance, [[LL10], Theorem 9.6]. We are going to prove that the ring condition (4.6) holds at any point x where $u(x)$ is finite. Without loss of generality, we prove it for $x = 0$.

For $z \in \mathbb{R}^N$ let

$$v_z(y) = \frac{\Gamma\left(\frac{N-2}{2}\right)}{2^2 \pi^{N/2}} |z - y|^{2-N} \quad \text{for all } y \in \mathbb{R}^N.$$

Then, v_z is superharmonic in $\mathbb{R}^N$ and $v_z(y) \to 0$ as $|y| \to \infty$.

By Lemma A.1 it follows that for all $z \in \mathbb{R}^N$, one has

$$f_R(z) := \fint_{B_R} v_z(y) dy \to 0 \quad \text{as } R \to \infty. \tag{A.2}$$

Using Fubini's theorem and the fact that $v_z(y) = v_y(z)$, we deduce

$$\fint_{B_R} u(z) dz = \fint_{B_R} \left(\int_{\mathbb{R}^N} v_z(y) d\mu(y) \right) dz$$

$$= \int_{\mathbb{R}^N} \left(\fint_{B_R} v_y(z) dz \right) d\mu(y)$$

$$= \int_{\mathbb{R}^N} f_R(y) d\mu(y),$$

for any $R > 0$. From (A.2), we have that $f_R \to 0$ pointwise as $R \to \infty$. Since v_z is superharmonic, we have

$$f_R(z) = \fint_{B_R} v_z(y) dy \leq v_z(0) = v_0(z) \quad \text{for any } R > 0,$$

and since $u(0) = \int v_0 d\mu < \infty$, we are in the position to apply the dominated convergence theorem concluding

$$\fint_{B_R} u(z)dz \to 0 \quad \text{as } R \to \infty,$$

that is, the ring condition (4.6) holds, and by Lemma A.1 we deduce that $\text{essinf}_{\mathbb{R}^N} u = 0$. $\square$

Appendix B
Harnack Inequalities for Quasilinear Elliptic Operators

In this section, we gather the main results on Harnack inequalities for quasilinear elliptic operators that were used in the current book. For convenience, we shall state them for m-Laplace operators and their weighted versions. The interested reader may consult Trudinger [Tru67] or Gilbarg and Trudinger [GT01] for a more general setting.

Let $\Omega \subset \mathbb{R}^N$ and consider the equation

$$\operatorname{div}\left(\nabla u|^{m-2}\nabla u\right) + g(x, u, \nabla u) = 0 \quad \text{in } \Omega, \tag{B.1}$$

where $m > 1$ and $g : \Omega \times \mathbb{R} \times \mathbb{R}^N \to \mathbb{R}$ is a measurable function such that

$$g(x, u, \eta) \leq a_0|\eta|^m + a_1(x)|\eta|^{m-1} + a_2(x)|u|^{m-1}, \tag{B.2}$$

with $a_0 \geq 0$, $a_1, a_2 \geq 0$ are measurable functions such that $|a_i(x)| \leq \mu$, $i = 1, 2$. By a weak solution of (B.1), we understand a function $u \in W_{loc}^{1,m}(\Omega)$ such that

$$\int_\Omega |\nabla u|^{m-2}\nabla u \cdot \nabla \varphi dx = \int_\Omega g(x, u, \nabla u)\varphi dx \quad \text{for all } \varphi \in C_c^\infty(\Omega). \tag{B.3}$$

Similarly, we say that $u \in W_{loc}^{1,m}(\Omega)$ is a weak subsolution (resp. weak supersolution) of (B.1) if (B.3) holds with the inequality sign $\geq$ (resp. the sign $\leq$).

Theorem B.1 (Strong Harnack Inequality, Trudinger [Tru67, Tru67]) *Let u be a weak solution of* (B.1) *in a ball $B_{3R}(z) \subset \Omega$ with $0 \leq u < M$ in $B_{3R}(z)$.*
Then, there exists $C = C(N, m, a_0 M, \mu R) > 0$ such that

$$\max_{\overline{B}_R(z)} u \leq C \min_{B_R(z)} u.$$

© The Author(s), under exclusive license to Springer Nature Switzerland AG 2022

M. Ghergu, *Partial Differential Inequalities with Nonlinear Convolution Terms*, SpringerBriefs in Mathematics, https://doi.org/10.1007/978-3-031-21856-9

Theorem B.2 (Weak Harnack Inequality, Trudinger [Tru67, Tru67]) *Let u be a weak subsolution of* (B.1) *in a ball* $B_{3R}(x) \subset \Omega$ *with* $0 \le u < M$ *in* $B_{3R}(z)$. *Then, for any* $p > m - 1$, *there exists* $C = C(N, m, p, a_0 M, \mu R) > 0$ *such that*

$$\max_{\overline{B}_R(z)} u \le C R^{-N/p} \|u\|_{L^p(B_{2R}(z))}.$$

The following results are simple consequences of Theorems B.1 and B.2, respectively.

Proposition B.3 (Strong Harnack Inequality) *Let* $u \in W_{loc}^{1,m}(\Omega) \cap C(\Omega)$, $u \ge 0$ *satisfy*

$$\mathrm{div}\big(|x|^{-\beta}|\nabla u|^{m-2}\nabla u\big) + a(x)u^{m-1} = 0 \quad \text{in } \Omega,$$

where $|a(x)| \le c|x|^{-m-\beta}$ *for some constant* $c > 0$. *Assume* $z \in \Omega$ *and* $R > 0$ *are such that* $B_{3R}(z) \subset \Omega$. *Then, there exists a constant* $C > 0$ *independent of* u *such that*

$$\max_{\overline{B}_R(z)} u \le C \min_{\overline{B}_R(z)} u. \tag{B.4}$$

Proof Note that u satisfies the equation

$$\mathrm{div}\big(|\nabla u|^{m-2}\nabla u\big) - \frac{\alpha}{|x|^2}|\nabla u|^{m-2}\nabla u \cdot x + b(x)u^{m-1} = 0 \quad \text{in } \Omega,$$

where $b(x) = a(x)|x|^\beta$ and $|b(x)| \le c|x|^{-m}$. The above equation fulfils the structural assumptions in Theorem B.1. According to this result, u satisfies (B.4).
                                                                                          □

Proposition B.4 (Weak Harnack Inequality) *Let* $R > 0$ *and* a, b, c *be real numbers such that* $a > b > 3c > 0$. *Assume* $\Omega \subset \mathbb{R}^N$ *is an open set such that*

$$\overline{B}_{(a+3c)R} \setminus B_{(b-3c)R} \subset \Omega.$$

Suppose $u \in W_{loc}^{1,m}(\Omega) \cap C(\Omega)$ *satisfies* $u \ge 0$ *and*

$$\mathrm{div}\big(|x|^{-\beta}|\nabla u|^{m-2}\nabla u\big) \ge 0 \quad \text{in } \Omega. \tag{B.5}$$

Then, for any $p > m - 1$, *there exists a constant* $C > 0$ *independent of* R *such that*

$$R^{N/p} \sup_{B_{aR} \setminus B_{bR}} u \le C \Big(\int_{B_{(a+2c)R} \setminus B_{(b-2c)R}} u^p \Big)^{1/p}. \tag{B.6}$$

Proof Observe first that (B.5) is equivalent to

$$\text{div}\big(|\nabla u|^{m-2}\nabla u\big) - \frac{\beta}{|x|^2}|\nabla u|^{m-2}\nabla u \cdot x \geq 0 \quad \text{in } \Omega,$$

which satisfies the structural assumptions in Theorem B.2.

Let $z_1, z_2, \ldots, z_k \in \Omega$ be such that $\{B_{cR}(z_i)\}_{1\leq i \leq k}$ is a finite cover with open balls of the compact set $\overline{B}_{aR} \setminus B_{bR}$. By Theorem B.2 we find

$$R^{N/p} \sup_{B_{cR}(z_i)} u \leq C\Big(\int_{B_{2cR}(z_i)} u^p\Big)^{1/p} \leq C\Big(\int_{B_{(a+2c)R}\setminus B_{(b-2c)R}} u^p\Big)^{1/p}.$$

Thus,

$$R^{N/p} \sup_{B_{aR}\setminus B_{bR}} u \leq R^{N/p} \sup_{\cup_{i=1}^{k} B_{cR}(z_i)} u \leq C\Big(\int_{B_{(a+2c)R}\setminus B_{(b-2c)R}} u^p\Big)^{1/p}.$$

$\square$

Bibliography

[AG01]   D.H. Armitage, S.J. Gardiner, *Classical Potential Theory*. Springer Monographs in Mathematics (2001)

[BCN94] H. Berestycki, I. Capuzo Dolcetta, L. Nirenberg, Superlinear indefinite elliptic problems and nonlinear Liouville theorems. Topol. Methods Nonlinear Anal. **4**, 59–78 (1994)

[BP01]   M.F. Bidaut-Véron, S. Pohozaev, Nonexistence results and estimates for some nonlinear elliptic problems. J. Anal. Math. **84**, 1–49 (2001)

[BM98]   I. Birindelli, E. Mitidieri, Liouville theorems for elliptic inequalities and applications. Proc. Roy. Soc. Edinb. Sect. A **128**, 1217–1247 (1998)

[BLU07] A. Bonfiglioli, E. Lanconelli, F. Uguzzoni, *Stratified Lie Groups and Potential Theory for Their Sub-Laplacians*. Springer Monographs in Mathematics (2007)

[BFP15] S. Bordoni, R. Filippucci, P. Pucci, Nonlinear elliptic inequalities with gradient terms on the Heisenberg group. Nonlinear Anal. **121**, 262–279 (2015)

[BCCT13] B. Brandolini, F. Chiacchio, F.C. Cirstea, C. Trombetti, Local behaviour of singular solutions for nonlinear elliptic equations in divergence form. Calc. Var. Partial Differ. Equ. **48**, 367–393 (2013)

[BV81]   H. Bré zis, L. Véron, Removable singularities for some nonlinear elliptic equations. Arch. Ration. Mech. Anal. **75**, 1–6 (1981)

[CDM08] G. Caristi, L. D'Ambrosio, E. Mitidieri, Liouville theorems for some nonlinear inequalities. Proc. Steklov Inst. Math. **260**, 90–111 (2008)

[CMP08] G. Caristi, E. Mitidieri, S. Pohozaev, Local estimates and Liouville theorems for a class of quasilinear inequalities (Russian). Dokl. Akad. Nauk **418**, 453–457 (2008); translation in Dokl. Math. **77**, 85–89 (2008)

[CZ16]   H. Chen, F. Zou, Classification of isolated singularities of positive solutions for Choquard equations. J. Differ. Equ. **261**, 6668–6698 (2016)

[CZ18]   H. Chen, F. Zou, Isolated singularities of positive solutions for Choquard equations in sublinear case. Commun. Contemp. Math. **20**, 1750040 (2018)

[DG22]   L. D'Ambrosio, M. Ghergu, Representation formulae for nonhomogeneous differential operators and applications to PDEs. J. Differ. Equ. **317**, 706–753 (2022)

[DMP06] L. D'Ambrosio, E. Mitidieri, S.I. Pohozaev, Representation formulae and inequalities for solutions of a class of second order partial differential equations. Trans. Am. Math. Soc. **358**, 893–910 (2006)

[DGK22] D. Devine, M. Ghergu, P. Karageorgis, Quasilinear elliptic inequalities with nonlinear convolution terms and potentials of slow decay. Differ. Integr. Equ. **36**(1/2), 1–20 (2023)

© The Author(s), under exclusive license to Springer Nature Switzerland AG 2022   131
M. Ghergu, *Partial Differential Inequalities with Nonlinear Convolution Terms*,
SpringerBriefs in Mathematics, https://doi.org/10.1007/978-3-031-21856-9

[DA10] J.T. Devreese, A.S. Alexandrov, *Advances in Polaron Physics*. Springer Series in Solid-State Sciences, vol. 159 (Springer, Berlin, 2010)

[Emd07] R. Emden, Gaskugeln: Anwendungen der Mechanischen Wärmetheorie auf kosmologische und meteorologische Probleme (Leipzig, B.G. Teubner, 1907)

[FG20] R. Filippucci, M. Ghergu, Singular solutions for coercive quasilinear elliptic inequalities with nonlocal terms. Nonlinear Anal. **197**, 111857 (2020)

[FG22] R. Filippucci, M. Ghergu, Higher order evolution inequalities with nonlinear convolution terms. Nonlinear Anal. **221**, 112881 (2022)

[FPR10] R. Filippucci, P. Pucci, M. Rigoli, Nonlinear weighted p-Laplacian elliptic inequalities with gradient terms. Commun. Contemp. Math. **12**, 501–535 (2010)

[Fow14] R.H. Fowler, The form near infinity of real continuous solutions of a certain differential equation of second order. Q. J. Math. **45**, 289–350 (1914)

[Fow20] R.H. Fowler, The solutions of Emden's and similar differential equations. Monthly Notices R. Astron. Soc. **91**, 63–91 (1920)

[FV86] A. Friedman, L. Véron, Singular solutions of some quasilinear elliptic equations. Arch. Ration. Mech. Anal. **96**, 359–387 (1986)

[Fuj66] H. Fujita, On the blowing up of solutions to the Cauchy problem for $u_t = \Delta u + u^{1+\alpha}$. J. Fac. Sci. Univ. Tokyo, Sect. 1A, Math. **13**, 109–124 (1966)

[GL98] V.A. Galaktionov, H.A. Levine, A general approach to Fujita exponents in nonlinear parabolic problems. Nonlinear Anal. **34**, 1005–1027 (1998)

[GT01] D. Gilbarg, N. Trudinger, *Elliptic Partial Differential Equations of Second Order* (Springer-Verlag Berlin, 2001)

[GKS20] M. Ghergu, P. Karageorgis, G. Singh, Positive solutions for quasilinear elliptic inequalities and systems with nonlocal terms. J. Differ. Equ. **268**, 6033–6066 (2020)

[GKS21] M. Ghergu, P. Karageorgis, G. Singh, Quasilinear elliptic inequalities with Hardy potential and nonlocal terms. Proc. R. Soc. Edinb. Sect. A **151**, 1075–1093 (2021)

[GMM21] M. Ghergu, Y. Miyamoto, V. Moroz, Polyharmonic inequalities with nonlocal terms. J. Differ. Equ. **296**, 799–821 (2021)

[GMT11] M. Ghergu, A. Moradifam, S.D. Taliaferro, Isolated singularities of polyharmonic inequalities. J. Funct. Anal. **261**, 660–680 (2011)

[GT15] M. Ghergu, S. Taliaferro, Asymptotic behavior at isolated singularities for solutions of nonlocal semilinear elliptic systems of inequalities. Calc. Var. Partial Differ. Equ. **54**, 1243–1273 (2015)

[GT17] M. Ghergu, S.D. Taliaferro, Pointwise bounds and blow-up for systems of semilinear parabolic inequalities and nonlinear heat potential estimates. J. Funct. Anal. **272**, 1301–1339 (2017)

[GT16a] M. Ghergu, S. Taliaferro, Pointwise bounds and blow-up for Choquard-Pekar inequalities at an isolated singularity. J. Differ. Equ. **261**, 189–217 (2016)

[GT16b] M. Ghergu, S. Taliaferro, *Isolated Singularities in Partial Differential Inequalities*. Encyclopedia of Mathematics and Its Applications, vol. 161 (Cambridge University Press, Cambridge, 2016)

[GS81] B. Gidas, J. Spruck, Global and local behavior of positive solutions of nonlinear elliptic equations. Commun. Pure Appl. Math. **34**, 525–598 (1981)

[Gue03] M. Guedda, Local and global nonexistence of solutions for a degenerate parabolic inequality. Appl. Math. Lett. **16**, 493–499 (2003)

[Har28a] D.R. Hartree, The wave mechanics of an atom with a non-Coulomb central field, Part I. Theory and Methods. Math. Proc. Camb. Philos. Soc. **24**, 89–110 (1928)

[Har28b] D.R. Hartree, The wave mechanics of an atom with a non-Coulomb central field, Part II. Some Results and Discussion. Math. Proc. Camb. Philos. Soc. **24**, 111–132 (1928)

[Har28c] D.R. Hartree, The wave mechanics of an atom with a non-Coulomb central field, Part III. Term Values and Intensities in Series in Optical Spectra. Math. Proc. Camb. Philos. Soc. **24**, 426–437 (1928)

[Hay73] K. Hayakawa, On nonexistence of global solutions of some semilinear parabolic differential equations. Proc. Jpn. Acad. **49**, 503–505 (1973)

[JS21]   M. Jleli, B. Samet, New blow-up phenomena for hyperbolic inequalities with combined nonlinearities. J. Math. Anal. Appl. **494**, 124444 (2021)

[JSY19] M. Jleli, B. Samet, D. Ye, Critical criteria of Fujita type for a system of inhomogeneous wave inequalities in exterior domains. J. Differ. Equ. **268**, 3035–3056 (2019)

[Jon95] K.R.W. Jones, Newtonian quantum gravity. Aust. J. Phys. **48**, 1055–1081 (1995)

[Kem72] J.T. Kemper, Temperatures in several variables: Kernel functions, representations, and parabolic boundary values. Trans. Am. Math. Soc. **167**, 243–262 (1972)

[KST77] K. Kobayashi, T. Sirao, H. Tanaka, On the blowing up problem for semi-linear heat equations. J. Math. Soc. Jpn. **29**, 407–424 (1977)

[KLM05] V. Kondratiev, V. Liskevich, V. Moroz, Positive solutions to superlinear second-order divergence type elliptic equations in cone-like domains. Ann. Inst. H. Poincaré Anal. Non Linéaire **22**, 25–43 (2005)

[KLZ03] V. Kondratiev, V. Liskevich, Z. Sobol, Second-order semilinear elliptic inequalities in exterior domains. J. Differ. Equ. **187**, 429–455 (2003)

[Lap02] G.G. Laptev, Nonexistence results for higher-order evolution partial differential inequalities. Proc. Am. Math. Soc. **131**, 415–423 (2003)

[Lap03] G.G. Laptev, Non-existence of global solutions for higher-order evolution inequalities in unbounded cone-like domains. Mosc. Math. J. **3**, 63–84 (2003)

[Li04]   Y.Y. Li, Remark on some conformally invariant integral equations: the method of moving spheres. J. Eur. Math. Soc. **6**, 153–180 (2004)

[LLM07] V. Liskevich, S. Lyakhova, V. Moroz, Positive solutions to nonlinear p-Laplace equations with Hardy potential in exterior domains. J. Differ. Equ. **232**, 212–252 (2007)

[Lie76] E.H. Lieb, Existence and uniqueness of the minimizing solution of Choquard's nonlinear equation. Stud. Appl. Math. **57**, 93–105 (1976/1977)

[LL10]   E.H. Lieb and M. Loss, Analysis, Second Edition, Amer. Math. Soc., 2010.

[Lio80] P.-L. Lions, The Choquard equation and related questions. Nonlinear Anal. **4**(25), 1063–1072 (1980)

[Lio84] P.-L. Lions, The concentration-compactness principle in the calculus of variations. The locally compact case. II, in *Ann. Inst. H. Poincaré Anal. Non Linéaire* (1984), pp. 223–283

[MP01]   E. Mitidieri, S.I. Pohozaev, A priori estimates and blow up of solutions to nonlinear partial differential equations. Proc. Steklov Inst. Math. **234**, 1–367 (2001)

[MNU02] M. Mizukami, M. Naito, H. Usami, Asymptotic behavior of solutions of a class of second order quasilinear ordinary differential equations. Hiroshima Math. J **32**, 51–78 (2002)

[MPT98] I.M. Moroz, R. Penrose, P. Tod, Spherically-symmetric solutions of the Schrödinger-Newton equations. Classical Quantum Gravity **15**, 2733–2742 (1998)

[MV13a] V. Moroz, J. Van Schaftingen, Nonexistence and optimal decay of supersolutions to Choquard equations in exterior domains. J. Differ. Equ. **254**, 3089–3145 (2013)

[MV13b] V. Moroz, J. Van Schaftingen, Groundstates of nonlinear Choquard equations: existence, qualitative properties and decay asymptotics. J. Funct. Anal. **265**, 153–184 (2013)

[MV07]   V. Moroz, J. Van Schaftingen, A guide to the Choquard equation. J. Fixed Point Theory Appl. **19**, 773–813 (2017)

[Pek54] S. Pekar, Untersuchung über die Elektronentheorie der Kristalle (Akademie Verlag, Berlin, 1954)

[Pen96] R. Penrose, On gravity's role in quantum state reduction. Gener. Relativ. Gravitation **28**, 581–600 (1996)

[PV00]   S.I. Pohozaev, L. Véron, Blowup results for hyperbolic inequalities. Ann. Scuola Norm. Sup. Pisa **29**, 393–420 (2000)

[PRS07] P. Pucci, M. Rigoli, J. Serrin, Qualitative properties for solutions of singular elliptic inequalities on complete manifolds. J. Differ. Equ. **234**, 507–543 (2007)

[PSZ99] P. Pucci, J. Serrin, H. Zou, A strong maximum principle and a compact support principle for singular elliptic inequalities. J. Math. Pures Appl. **78**, 769–789 (1999)

[QS07]   P. Quittner, P. Souplet, *Superlinear Parabolic Problems. Blow-up, Global Existence and Steady States*. Birkhauser Advanced Texts (2007)

[R30]   F. Riesz, Sur les fonctions sous harmoniques et leur rapport á la theorie du potentiel II, Acta Math. 54 (1930), 321–360.

[SYW16] H. Song, J. Yin, Z. Wang, Isolated singularities of positive solutions to the weighted *p*-Laplacian. Calc. Var. Partial Differ. Equ. **55**, 55–28 (2016)

[Tal07] S.D. Taliaferro, Isolated singularities of nonlinear parabolic inequalities. Math. Ann. **338**, 555–586 (2007)

[Tru67] N. Trudinger, On Harnack type inequalities and their application to quasilinear equations. Commun. Pure Appl. Math. **20**, 721–747 (1967)

[VV80]   J.L. Vázquez, L. Véron, Removable singularities of some strongly nonlinear elliptic equations. Manuscripta Math. **33**, 129–144 (1980)

[Ver81] L. Véron, Singular solutions of some nonlinear elliptic equations. Nonlinear Anal. **5**, 225–242 (1981)

[Wat76] N.A. Watson, Green functions, potentials, and the Dirichlet problem for the heat equation. Proc. Lond. Math. Soc. **33**, 251–298 (1976)

Index

© The Author(s), under exclusive license to Springer Nature Switzerland AG 2022   135
M. Ghergu, *Partial Differential Inequalities with Nonlinear Convolution Terms*,
SpringerBriefs in Mathematics, https://doi.org/10.1007/978-3-031-21856-9

Printed in the United States
by Baker & Taylor Publisher Services

Printed in the United States
by Baker & Taylor Publisher Services